STRUCK BY LIGHTNING

STRUCK BY LIGHTNING

THE CURIOUS WORLD OF PROBABILITIES

JEFFREY S. ROSENTHAL

Joseph Henry Press
Washington, D.C.

Joseph Henry Press • 500 Fifth Street, NW • Washington, DC 20001

The Joseph Henry Press, an imprint of the National Academy Press, was created with the goal of making books on science, technology, and health more widely available to professionals and the public. Joseph Henry was one of the founders of the National Academy of Sciences and a leader in early American science.

Library of Congress Cataloging-in-Publication Data

Rosenthal, Jeffrey Seth.
Struck by lightning : the curious world of probabilities / Jeffrey S. Rosenthal.
p. cm.
Includes index.
ISBN 0-309-09734-7 (hardback)
1. Probabilities. 2. Random variables. 3. Chance. 4. Stochastic processes. I. Title.
QA273.R785 2006
519.2—dc22

2005037021

Published by arrangement with HarperCollins Publishers Ltd., Toronto, Canada.

Printed in the United States of America.

Contents

1

Surrounded by Randomness

Probabilities Everywhere

When I was a graduate student at Harvard University, I booked a flight to New York's John F. Kennedy Airport to visit some relatives. Exactly one week before my flight, there was a major accident at the same airport: an Avianca airplane missed its first landing approach, ran out of fuel during its second approach, and crashed, killing 73 people.

At first I was shocked. How could I travel to Kennedy airport so soon after such a tragedy? Surely it wasn't safe. I would have to cancel my visit!

In an effort to calm myself, I tried to think logically. At the time, I was working on a doctoral dissertation related to the mathematics of probability theory, but my research was rather theoretical and not related to everyday life. Could I perhaps apply my abstract knowledge to this very concrete situation?

I did a quick calculation. I determined that there are about 5,000 flights per week to Kennedy airport. So, even if this airport accident was somehow the fault of the Kennedy Airport itself (which it probably wasn't), and even if I knew that there would be another accident there sometime during the next week (which I didn't), there would still be only one chance in 5,000 that my own flight would be affected.

Now, one chance in 5,000 isn't so small, but it isn't so large either. It was small enough to reassure me that my flight would probably be fine.

I flew to New York as scheduled and, in a victory for probability theory, I encountered no difficulties whatsoever.

In the years following my New York adventure, I began to realize that we are all constantly faced with situations and choices that involve randomness and uncertainty. A basic understanding of the rules of probability theory as they apply to real-life circumstances can help us to make sense of these situations, avoid unnecessary fear, seize the opportunities that randomness presents to us, and actually enjoy the uncertainties we face.

Human beings have always been both fascinated and appalled by randomness. On the one hand, we love the thrill of a surprise party, the unpredictability of a budding romance, the mystery of a good detective novel, and the freedom of not knowing what tomorrow may bring. We are inexplicably delighted by strange coincidences and striking similarities. Staid urban sophisticates gladly plunk down millions of dollars on lottery tickets and horse races—and on the stock market. Weary adults, after a hard day of work, are happy to play cards or roll the dice in a feisty game of chance. Audiences root for *Casablanca*'s inconsistent and unpredictable Rick Blaine (Humphrey Bogart), over the brave and heroic—but predictable—Victor Laszlo (Paul Henreid).

On the other hand, we hate uncertainty's dark side. From cancer to SARS, diseases strike with no apparent pattern, ruining lives and baffling medical science. Terrorists attack, airplanes crash, bridges collapse, and we never know who will die next. Even the weather can suddenly and unexpectedly turn deadly, or just ruin an outdoor wedding. Successful politicians usually go out of their way to sound sure about everything, thus helping us to forget their (and our) lack of control over major national events that are ultimately, well, random.

Randomness is often neither good nor bad, but simply confusing. We are told of poll results accurate "to within four percentage points, 19 times out of 20," or of a study that "proves" that a certain medication is good and certain lifestyle choices are bad. We are told that we have "nothing to lose" when we nervously ask someone out on a date, despite the frightening possibility of rejection. We are assured that there is a "40% chance of rain today." We are warned of the risk of "false positive" results when we are diagnosed with a disease. It is not always clear how we should react to such probabilities, or even what they really mean.

At times we try to ignore or explain away the random nature of our world. We imagine that God creates the weather intentionally, to punish us, or that counting flower petals can tell us whether "she loves me or she loves me not." In Shakespeare's *Julius Caesar,* Cassius denies the effect of luck on human destiny, declaring, "The fault, dear Brutus, is not in our stars/But in ourselves, that we are underlings." In the movies, poker-playing, cigar-chewing cowboys get Royal Flushes through sheer willpower. In *The Cooler,* Bernie (William H. Macy) causes gamblers to lose just by standing nearby. And the rascally hero of *Star Wars,* Han Solo (Harrison Ford), warned by an android that the odds against successfully navigating an asteroid field are approximately 3,720 to one, scoffs "Never tell me the odds!"

But the reality is that when it comes to randomness, you can run but you can't hide. Many aspects of our lives are governed by events not completely in our control, and uncertainty is here to stay. We have two options: we can let uncertainty get the better of us or we can learn to understand randomness. If we do the latter, we will make better choices and learn to harness uncertainty for our own purposes.

Let's consider a few scenarios and how an understanding of uncertainty and probability might help us sort them out.

- *You are planning a trip to a foreign country, but you are put off by reports of terrorist activities there. Should you go anyway?* A simple understanding of probabilities will allow you to assess the risk of terrorism on your trip, and to decide whether the probabilities are high enough to affect your plans.
- *You need a secret code to conduct a secure financial transaction over the Internet.* If you just make up a code, an enemy agent might anticipate your psychology, guess the code, and obtain secret information. On the other hand, if you use randomness to generate your

code, you can virtually guarantee security against even the cleverest enemy. Modern computers use randomness in this way all the time.

- *You are involved in a battle of wits with a clever opponent and want to avoid being outsmarted.* You can use randomness to create a Nash equilibrium strategy in which your opponent can do no better than guess.
- *The local police chief and politicians all insist that crime is out of control and that more money is needed for law enforcement.* You can use linear regression to decide for yourself whether or not crime is actually increasing.
- *You are considering asking out that cute accountant in the business office, but you are worried that she will reject your advance or perhaps even complain about it.* Utility theory allows you to quantify your feelings about your wants and fears, and to compute whether or not they justify making that phone call.
- *Your doctor informs you that you must take a certain drug, which is shown to be effective by the latest medical study.* By considering the study's biases and p-value, you can determine for yourself whether or not to accept its conclusions.
- *A rival scoffs that you are more likely to be struck by lightning than to succeed in your business venture.* A simple check of the numbers will reveal just how unlikely lightning deaths really are and put your rival's comments into context.
- *You are annoyed by all the spam e-mail you receive, and you wish you could find a way to block it.* Probability theory helps computers to separate spam from genuine messages, sparing you the burden of a jammed mailbox.
- *In one day, you see three different people who have all dyed their hair green. Is this a new and popular fashion trend?* Random events tend to occur in bunches due to Poisson clumping; many apparently striking coincidences or trends are the result of pure chance, and are of no meaning or consequence.
- *A friend tries to stump you with the "Monty Hall problem": which of three doors is most likely to have a car behind it, if you already*

know that Door #3 is empty? The theory of conditional probability allows you to compute all the odds and make the right choice.

- *You write a fantastic song, but you worry that perhaps someone else may have already written the exact same song.* Probability theory provides a useful perspective on uniqueness and practically guarantees that your song is brand new.
- *You wonder how scientists and engineers compute all of the complicated quantities required to build bridges, conduct medical studies, and design nuclear reactors.* Monte Carlo sampling uses randomness on high-speed computers to compute many such quantities.
- *You need to decide whether to call a poker bet or how many houses to buy in Monopoly.* Probability theory provides many insights into strategies for games of chance, and using it allows you to win more often in the long run.

These scenarios are many and varied, but they all have one thing in common. In each case, knowing the rules of probability, randomness, and uncertainty allows us to make better decisions and to understand the world around us more clearly. Even simple probability calculations can help reduce our stress, and clarify our choices, by putting randomness in perspective: a "Probability Perspective" based on rational thought about randomness rather than on irrational emotional responses.

While no one can predict uncertain events with certainty, we can at least understand the uncertainty itself. This book will discuss the probabilities associated with many different events. By thinking logically about the likelihood of various outcomes, we can make better decisions and understand our lives more deeply. We can better cope with the uncertainties we face and perhaps even learn to enjoy them.

So the next time your daughter is flying in for the holidays, in the midst of a rainstorm and with thunder and lightning across the skies, don't panic. Don't despair. Don't fill your mind with images of horrible accidents. Instead, remember the Probability Perspective. Remember that each year there are about 10 million commercial flights in the United States alone—including many during rainstorms—and on average only about five crashes involving fatalities. The probability that your daughter's flight will result in even one death is only about one chance in 2 million. It just isn't going to happen.

In place of worrying, enjoy the moment. Eagerly anticipate her visit. Cook her favorite meal. Prepare for a fun game of chance involving cards or dice. Think about the fascination and intrigue that randomness adds to our daily lives.

And when your daughter finally arrives, a little wet and very hungry but perfectly safe, be sure to give her a big hug.

2

What Are the Odds of That?

Coincidence and Surprise

We are often struck by seemingly astounding coincidences. You meet three friends for dinner and discover that all four of you are wearing dresses of the same color. You dream about your grandson the day before he phones you out of the blue. Your two office mates both get called for jury duty on the same day. You discover that your boss's new bride went to the same tiny elementary school that you did. Such events amuse us, or fascinate us, or arouse our suspicions, or evoke deep mystical significance. But should they?

From the Probability Perspective, our first question should be, how unlikely was the event? Was the coincidence run-of-the-mill or truly astounding?

Out of How Many?

Everything that happens is surprising in some sense, from someone's perspective.

The Astounding Lottery Winner

"I don't believe it," Jennifer exclaims. "John Smith from Smalltown just won the lottery jackpot!"

"Wow, that's great," you reply cautiously. "Do you know him?"

"No, unfortunately not."

"Had you heard of him before?"

"No, never."

"Have you ever been to Smalltown?"

"Nope."

"So why are you so surprised?"

"Because the probability of winning that lottery jackpot is about one in 14 million," Jennifer declares with an air of authority. "And yet John Smith pulled it off!"

Winning any commercial lottery jackpot is extremely unlikely. On the other hand, millions of people buy lottery tickets every day, and usually at least one of them will win. This doesn't surprise us at all, but why not? The reason is that one person out of millions has won the lottery. There are millions of different chances for someone to win the lottery, so of course someone usually does.

If you flip a coin 10 times and get heads every time, that would be rather surprising, because the chances of its happening were 1 in 1,024 (computed by taking a factor of 1/2 for each of the ten coins and multiplying them all together), which is less than 0.1%. However, if you spend an entire afternoon repeatedly flipping the same coin, and after several hours you finally get 10 heads in a row, that was bound to happen and is not surprising at all.

So, whenever a friend announces a surprising development, the first thing you should ask yourself is, out of how many? That is, how many different chances were there for that event—or any other similarly surprising one—to arise?

A Cousin at Disney World (A True Story)

When I was 14 years old, my family traveled to Orlando, Florida, to visit Disney World. For two days we went on scary roller coasters and gentle train rides, saw haunted houses and singing puppets, and ate lots of junk food. Remarkably, in the midst of thousands of strangers, we ran into my father's cousin Phil and his family. They lived in Connecticut, and none of us had any idea that the other family was in Florida. We were all stunned at the coincidence.

How surprised should we have been? There were about 230 million people in the United States at that time. So, the probability that any one person chosen at random at Disney World would be my father's cousin Phil, was something like one chance in 230 million, unimaginably low. However, over the course of our two days at Disney World, we had passed many different strangers waiting in line for many different rides and treats. In total there must have been *at least* 2,000 people that we saw close enough to recognize, any one of whom could have been Phil. Right away this increases the probability by a factor of 2,000, to one chance in 115,000.

But Cousin Phil wasn't the only person we might have run into. What about my father's other cousins? What about my mother's cousins? What about numerous other relatives? Or our friends or work colleagues? Or our neighbors? Or classmates? Or relatives of friends? Or friends of neighbors? There must be at least 500 people whom we could have run into who would have surprised us as much as Phil did. This increases the probability by another factor of 500, to one chance in 230.

Of course, one chance in 230 is still less than half of one percent. Therefore, on most trips to Disney World, you probably won't run into anyone that you know. But still, over a lifetime of traveling and visiting and exploring, you are bound to run into people unexpectedly, now and then. It really isn't so surprising after all.

This "out of how many" issue comes up in many ways. For example, a friend told me that the night before her father died, she had a dream in which her father appeared, looking surprisingly peaceful. Some might consider that this dream showed that my friend somehow "knew" that her father was about to die, or even that my friend's father had somehow communicated with her on a subconscious level over the 500 kilometers between them.

Perhaps. But another explanation is that we all dream about many things every night. The dreams that we are most likely to remember or take note of or discuss with others are those that happen to have some surprising connection to other events. My friend might dream of her father on perhaps one night in 50, so the probability that she would dream about him the night before he died is only about one in 50. However, the probability that she would have some dream at some point in her life that would have some connection to some event, is much much higher. Thus, the question is: one profound dream out of how many dreams in total?

Nobel Prize winner Richard Feynman wrote about an incident that occurred when he was a student. Suddenly, the physicist got a feeling that he *knew,* somehow, that his grandmother had died. Just then the phone rang. Had his prediction come true? Had his grandmother passed away? No, the phone call was for another student, and Feynman's grandmother was just fine. This story nicely illustrates that very often we have hunches or dreams or predictions, but we tend to forget the ones that don't come true. Then, when one does come true, we forget that this is one hunch out of many, so the observation is less surprising than it may appear.

Consider this classic physics question. Suppose you take a cup of water and pour it into the ocean. Gradually, currents and tides and rain and evaporation mix all of the world's water together. Five years later, you go to some other ocean, on the other side of the world, and fill a fresh glass with water. How many molecules of water from the first glass will end up in the second?

The world's oceans contain a lot of water, about a billion cubic kilometers' worth. Compared with that volume, one cup of water is hardly anything: about two parts in the number written as 1 followed by 22 zeros. So, the probability of any one particular molecule from your first glass making its way to your second glass is just two chances in 1 followed by 22 zeros: that is, practically impossible.

On the other hand, molecules are unimaginably tiny, and even one cup of water contains a tremendous number of them, almost as many as 1 followed by 25 zeros. Indeed, there are so many molecules in that first small glass of water that, purely by chance, over 1,000 of them will arrive five years later in the second glass of water. Now, 1,000 molecules sounds like a lot but, again, out of how many?

Can we apply similarly rational probability logic to the magic of love? Many people have a story of just "happening" to meet their future spouse under very unlikely circumstances. In my own case, while my wife, Margaret, and I had previously met briefly at a party, we truly interacted only when I just *happened* to walk to a post office one afternoon, accompanying a friend who just *happened* to be assisting a colleague by forwarding some packages on his behalf. Meanwhile, Margaret just *happened* to get off work early that day, and just *happened* to require postal services herself, and just *happened* to choose the exact same post office (which wasn't near her home or work).

What are the odds that I would accompany my friend on this mission? What are the chances that Margaret would go to the very same post office on the very same day? What is the probability that we would be there at the same exact time? Multiplying all these factors together, surely the probability of our meeting up that day was one chance in tens of thousands, at most.

And yet it happened. How do I explain that? Well, I *could* argue that even if Margaret and I hadn't met at the post office, we might have met somewhere else, later on, perhaps at another party. Or I could reason that of all the errands I run each week, something interesting was bound to happen some time. Or I could declare dispassionately that each person actually has many potential compatible partners and that sooner or later you will eventually meet one of them. But none of these arguments seems too convincing. I guess I'm just a lucky guy.

Friends of Friends of Friends

Sometimes we discover seemingly surprising chains of connections, like finding out that your next-door neighbor is your brother's housekeeper's cousin. When should we be surprised and when shouldn't we?

Some chains are more interesting than others. When the evil *Star Wars* character Darth Vader declares to the heroic Luke Skywalker, "I am your father," the revelation is unexpected and profound. But in Mel Brooks's spoof version, *Spaceballs,* when Lord Dark Helmet instead says "I am your father's brother's nephew's cousin's former roommate," the observation is far more mundane (and therefore an amusing parody). Why, precisely, do we perceive these two similar-sounding facts so differently?

Again, the explanation is, out of how many? We each have just one father. While there may be a few others of equal importance to us—mother, siblings, children, perhaps lifelong best friend—the position of father is still quite an exalted one, surely a "top 10" close association. As a result, it is quite surprising if our arch-enemy turns out to be so closely related to us. On the other hand, there are many, many people who are as important to us as our "father's brother's nephew's cousin's former roommate." So the fact that Lord Dark Helmet occupies such a distant role merely places him as one out of a tremendous number of associates. There is no reason to be surprised at this, and we are not. But just how many such associates are there? How many people are there who are related to you by some short chain of relations like this?

This question is related to the *six degrees of separation* phenomenon. This phrase originated in an experiment performed by the Harvard psychologist Stanley Milgram back in 1967. He mailed out packages to randomly selected people in Kansas and Nebraska, with instructions to try to forward the package to a particular "target" individual in Massachusetts. The catch was that recipients were allowed to forward the package only to someone they knew on a first-name basis. Thus, a recipient in Kansas would have to think: Whom do I know on a first-name basis who might in turn know someone who knows someone who knows this Massachusetts target?

Milgram found that, of the packages that were returned, the average number of steps in the chain was about six. Hence, the concept of "six degrees of separation" was born, with the intriguing idea that we are all connected to each other by short chains of "friends of friends of friends" or "associates of associates of associates."

Milgram's experiment had many flaws. It was restricted to just one country. And many packages never arrived, perhaps indicating simple indifference but perhaps also suggesting that longer chains were necessary (and never completed) in those cases. On the other hand, the recipients had to decide for themselves to whom they should forward the package, and they had no way of knowing which choice would lead to the shortest chain. Overall, most scientists accept Milgram's concept that we are all connected through relatively short chains of acquaintances, though the precise figure of six is perhaps more apocryphal than real.

There is another way to think about these connections. Suppose each person has, say, a total of 500 "friends," that is, people they know on a first-name basis. Then the number of friends-of-friends chains is 500 × 500, or 250,000. The number of friends-of-friends-of-friends chains is equal to 500 × 500 × 500, or 125 million. Of course, some of these different chains will lead back to the same person, so your total number of friends-of-friends-of-friends is somewhat less than 125 million, but it is still awfully large. So, six degrees of separation, even for the entire world's population, seems pretty reasonable.

Mathematicians love such chains of connections, and they have their own unique version, which is based on the work of the brilliant Hungarian mathematician Paul Erdös (1913–96). Erdös spent his life living out of a suitcase while visiting other mathematicians around the world. The mathematicians would take care of his daily needs for the privilege of collaborating with him to solve their research problems. As a result, Erdös was a co-author of over 1,500 published papers, written jointly with hundreds of different mathematicians from around the world.

Inspired by this extensive web of connection, mathematicians invented *Erdös numbers*. Every mathematician who ever wrote a paper jointly with Paul Erdös gets an Erdös number of 1 (there are over 500 such people). Everyone who ever wrote a paper jointly with someone who has an Erdös number of 1 in turn gets an Erdös number of 2 (there are about 7,000 such people). And so on. My own Erdös number is 3: I published a paper in 1999 with the mathematician Robin Pemantle, who in turn had published a paper in 1996 with the mathematician Svante Janson, who himself published a paper with Paul Erdös that same year—thus providing a chain of three mathematicians from me to Erdös. Unfortunately, this is not such a special accomplishment; over 33,000 other mathematicians also have an Erdös number of 3.

Not to be left out, movie lovers have created their own version of Erdös numbers; these are known as *Bacon numbers*. Everyone who has ever co-starred in a movie with actor Kevin Bacon gets a Bacon number of 1; everyone who has co-starred in a movie with one of those people gets a Bacon number of 2, and so on. For example, Susan Sarandon has a Bacon number of 2, because she co-starred in *Dead Man Walking* with Sean Penn, who in turn co-starred in *Mystic River* with Kevin Bacon.

Similar ideas have been applied to such diverse items as World Wide Web links (how many hyperlinks do you have to click on, with your Web browser, to travel a path from your Web page to mine?); to rock musicians who recorded music together (for example, Ray Charles is just three steps away from Ozzy Osbourne, because Charles recorded with Michael Jackson, who recorded with the heavy-metal guitarist Slash, who in turn recorded with Osbourne); to baseball players who have been teammates (for example, there are several different paths, each five players long, that link 1930s slugger Babe Ruth to current pitching star Roger Clemens); to Marilyn Monroe numbers (for which decorum prevents me from specifying what type of connection is counted).

It seems that when it comes to finding different sorts of connections between different sorts of folks, the sky is the limit. This puts "coincidences" in a whole new perspective: there are so many possible connections, that it is no surprise that they occasionally coincide by pure chance.

Counting the Pairs

One fun probability fact is known as the *birthday problem*. It says that if 23 people are selected at random, there is just over a 50% chance that two of them will have the same birthday (meaning day and month only, not year). With 41 or more people, the probability is over 90%. This makes for a nice parlor trick; the next time you're at a decent-sized party, find a poor sucker who hasn't yet heard of the birthday problem, bet him that two people in the room have the same birthday, and watch the money roll in.

Why are these probabilities so high? Once again, the question is, out of how many? But this time, the answer is a bit more subtle. Our first reaction is to say that there are 365 days in a year (let's not bother with leap years), so with just 23 people, we have covered only 23 out of 365 days, or 6.3%. This is a very small probability, which casts doubt on the validity of our parlor trick. However, this is the right answer to the wrong question. If you asked 23 people if any of them had their birthday *today* (or Christmas Day, or any other one specific day), there would indeed be only a 6.3% chance that one of them would answer yes.

(This fact once helped me with some detective work. At a statistics conference dinner, three people's birthdays were being celebrated. I was suspicious; there were only 180 people in the room, so statistically there should have been only 180 divided by 365, or about one half of a person, who had a birthday on that particular day. I was right to wonder; it later turned out that two of those "birthday boys" actually had their birthdays the following month and were just cashing in on the celebration.)

But with our parlor trick, we are asking whether any of 23 randomly selected people have the same birthday as anyone else's birthday, not a specific day like today or Christmas. And that difference explains why the probability is so much higher.

The reason is that there are a lot more *pairs* of people than there are people. For example, suppose there are just four people at the party: Amy, Betty, Cindy, and Debbie. Then the number of pairs of people is six: Amy–Betty, Amy–Cindy, Amy–Debbie, Betty–Cindy, Betty–Debbie, and

Cindy–Debbie. The more people at the party, the more pairs—a lot more. With 23 party-goers there are 253 pairs, while with 41 party-goers there are 820 pairs. (The number of pairs is equal to the number of party-goers, multiplied by 1 less than that number, divided by 2. So, with four people, there are $4 \times 3 \div 2 = 6$ pairs; with 23 people, there are $23 \times 22 \div 2 = 253$ pairs; with 41 people, $41 \times 40 \div 2 = 820$ pairs.)

Now we can see why the birthday problem works. Even with only 23 people, there are 253 *pairs* of people. Any one of those pairs has probability 1/365 of having the same birthday, so the average number of pairs with the same birthday is 253/365, which equals 0.69 of a pair. Since 0.69 is well over 0.5, this suggests better than a 50% chance of a birthday match. The figure 0.69 represents a little bit of overcounting, since it's possible that several different pairs will all have the same birthday. Correcting for this, it turns out that the true probability of at least one matching birthday out of 23 people works out to 50.7%. At a party of 41 people, the average number of pairs with the same birthday is 820/365, or 2.25 pairs, and the probability of at least one pair is a whopping 90.3%.

Table 2.1 Probabilities and Average Numbers of Birthday Matchings

# People	# Pairs	Average # Matches	Prob. of Match
4	6	0.02	1.64%
10	45	0.12	11.69%
20	190	0.52	41.14%
23	253	0.69	50.73%
30	435	1.19	70.63%
35	595	1.63	81.44%
40	780	2.14	89.12%
41	820	2.25	90.32%
45	990	2.71	94.10%
50	1,225	3.36	97.04%

The point is that probabilities of matchings—whether birthdays, or incomes, or hometowns, or favorite novels, or the amount of change in pockets, or anything else—are much higher than you might think, because there are so many possible pairs out there. So the next time two different items happen to match up, don't jump out of your skin. Instead, ask yourself, out of how many pairs?

Musical Mayhem

You're excited because you've just purchased the Super Special Song Spooler digital music device. Eagerly you download 4,000 of your favorite songs, and press the button to play them in random order. The headphones pulsate as guitar riffs mingle with drum solos in a scintillating symphony of sounds. You look forward to musical variety and bliss for the rest of your days.

As the 75th song begins, you are shocked to hear that it is a repeat of the 42nd song. In choosing just 75 songs out of 4,000, that stupid music device has already chosen a repeat. How could this be? The device must be defective!

Before demanding a refund, you do some calculations. Out of a collection of 75 songs, the number of *pairs* of songs is equal to 2,775. That's a lot of pairs, more than half of 4,000, the total number of songs available. In fact, when choosing 75 songs at random, out of the 4,000 available, the probability is 50.2% that you will hear at least one repeat.

So, perhaps the device isn't to blame after all. You decide to keep your headphones on.

When It Rains, It Pours

In the first week of November 2003, five separate homicides were recorded in the Greater Toronto metropolitan region, an area that averages just 1.5 homicides per week. This fact was widely reported in the media, amid fears of a huge and increasing crime wave. Toronto's police chief called for a public inquiry into the judicial system, saying it "provides no apparent deterrent." Was such concern justified?

Before answering that question, consider the following puzzle. Figure 2.1 shows two different sets of 100 dots each. One set of dots was placed purely at random, with each dot equally likely to appear anywhere in the box. The other set of dots was placed more deliberately. Which set of dots is truly random: the left set or the right set?

Figure 2.1 Which Dots Are Truly Random – the Left or the Right?

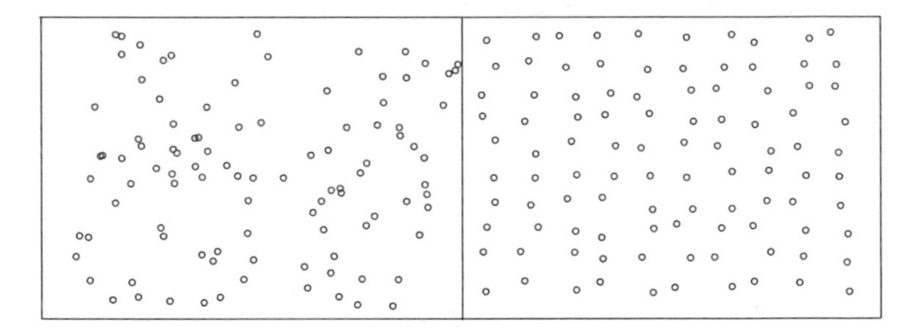

In this diagram, the dots on the left side exhibit various patterns. In some places, two or three dots are very close together. In other places, there are large areas with only a few dots. In fact, the dots toward the bottom almost appear to be following a spiral pattern; they don't appear random at all.

By contrast, the dots on the right are nicely spread out. No dot is too close or too far from another. No region is too sparsely covered. These dots seem to be nicely random. However, it is the dots in the left set that are truly random. In that set, each dot was placed anywhere in the box with equal probability, without regard to any other dot. All of the nearby dots, sparse regions, spiral patterns, and so on, happened purely by chance (I promise). It is the set of dots on the right that are not truly random. For those dots, I started with an evenly spread grid of dots and just applied a very small random movement to each dot (so they wouldn't line up perfectly). The dots in the right half have just a *tiny* bit of randomness (enough to fool the eye) but are still carefully designed to be distributed evenly over the entire box in a highly non-random way.

This example illustrates that wherever there are lots of events (like dots), some of them will tend to clump up together here and there, purely by chance, without signifying anything at all about trends or causes or connections. Randomness tricks us into seeing patterns and relationships that are really nothing more than chance occurrences. Probabilists refer to this phenomenon as *Poisson clumping*. The precise probabilities (the *Poisson distribution*) were first computed by the French mathematician Siméon-Denis Poisson in 1837.

Poisson clumping can make some people jumpy. I was once hired as a consultant to an on-line gambling Web site, because the manager thought that the computer wasn't selecting keno balls correctly. The balls should have been selected uniformly at random, but the manager thought that too many of them were coming from positions on the display board that were close to the number 17, and he feared that a sly customer could use this fact to make a profit. He needn't have worried; a careful statistical analysis confirmed that the balls were indeed being chosen with uniform probabilities. What the manager had seen was nothing more than Poisson clumping at work.

So what about those Toronto homicides? The Poisson distribution tells us that with an average of 1.5 homicides per week, there is still a 1.4% probability of seeing five homicides in a given week, purely by chance. So, we should expect to see five homicides in the same week once every 71 weeks—nearly once a year! It really isn't so surprising, nor does it signify anything other than bad luck.

By contrast, each week has just over 22% probability of being completely homicide-free, and indeed Toronto experiences many weeks without any homicides at all. But I have yet to see a newspaper headline that screams "No Murders This Week!"

Another way to think about Poisson clumping is in terms of pairs and larger groups. In the diagram above, there are 100 dots in the left-hand box. This means there are nearly 5,000 pairs of dots. With so many pairs, it is not surprising that some of the pairs are in very close proximity. Similarly, with 78 homicides in a year, there are 3,003 pairs of homicides, and over *21 million* different groups of five homicides. With so many five-homicide groups to choose from, it is not surprising that some group of five homicides would all take place within the same week.

Poisson clumping also explains why you sometimes get invited to two or three parties on the same night, after several party-free weeks. Or encounter two bad drivers within a few minutes, after several hours of problem-free driving. Or get treated rudely by five different people in one day. Or get three wrong-number telephone calls on the same evening. Or see three buses arriving in a row for no reason, after 25 frustrating minutes without a bus. There is nothing mystical or alarming or unexpected about these events. They are just a consequence of the laws of probability.

Speaking of buses, Poisson clumping also explains the importance of bus schedules. If buses arrive regularly, once every 10 minutes, then you have to wait between zero and 10 minutes for the bus, which works out to five minutes on average. However, if the buses arrive completely independently, without regard for the schedule or each other, this is no longer the case. You might be lucky and see a bus (or even several buses) right away, or you might be unlucky and have to wait a long time. It turns out that on average you will have to wait 10 minutes for a bus in this case—exactly *twice* as long as you would have if the buses were following a schedule. So, if your city's buses are passing by in a haphazard manner, complain to the mayor that your average waiting time has been unnecessarily doubled.

We see that many so-called coincidences can be explained as pure chance. The seemingly surprising observation may be just one out of many possibilities (like seeing my dad's cousin at Disney World), or be caused by the large number of pairs (like the birthday problem), or be due to Poisson clumping (like the weekly homicide count, or the spiral pattern of the dots).

Other times, coincidences arise because two seemingly unrelated events actually have the same cause. Bank profits shoot up just when housing prices tumble; both phenomena are due to rising interest rates. Or airline traffic and museum attendance both skyrocket in the same week, which happens to be spring break.

Sometimes such *common causes* are very subtle, and go unnoticed—even though they explain seemingly surprising coincidences. The next time two or more people suddenly make the same suggestion, or discuss the same problem, or take the same action, ask yourself whether some common cause might explain the surprise.

Parental Musings

You finally found a babysitter for Jason, and you and your husband are enjoying a rare moment of quiet romance at an elegant French restaurant.

As you luxuriously swirl your $15 glass of red wine, a few drops spill onto the white tablecloth. The tiny patches of color remind you of the craziness of Jason's fifth birthday party last year, when more spaghetti sauce ended up on the carpet than in the children's stomachs.

You remember how happy Jason seemed that day. A month later, he started kindergarten, and since then he has seemed more tense. When you discussed this change with his teacher in the fall, she agreed that Jason had had trouble adjusting to the new routines, but said that he seemed to be improving.

At the kindergarten music show a few weeks later, Jason smiled a lot and he enjoyed the event. He sat beside little Haldor, his new best friend. Haldor seems like a good kid.

Haldor, whose parents are Norwegian, had sung a traditional Norwegian folk song. He wasn't a great singer, but the song still evoked lovely images of spectacular fjords and happy people. Scandinavia would be an interesting region to visit some day.

"Maybe we should take a vacation in Scandinavia," your husband suddenly declares. You stare at him, stunned. "Why, I was just thinking about Scandinavia myself!" you exclaim. "What an astounding coincidence!"

Unbeknownst to you, your husband had also noticed those drops of red wine. He, too, had been reminded of Jason's party, and the kindergarten show, and Jason's new friend Haldor, and . . .

Even if there is no common cause, most coincidences can be explained by some combination of the "out of how many" principle, and Poisson clumping, and the multiplicative powers of "friends of friends." That doesn't make coincidences any less striking, but it does help us to understand why they occur so frequently.

3

Laying Down the Law

Why Casinos Always Win

Whether in the opulence of Las Vegas, or in the thousands of more modest operations around the world, or even just in the movies, we have all seen casinos at work. These establishments are filled with anxious gamblers and curious tourists who are placing bets on everything from slot machines, to roulette wheels, to crap shoots, to keno draws, to poker and blackjack tables. On any given bet, anything can happen: some players lose, others win. Some get rich, others go broke. Some stay and bet all day, others leave after just a few minutes.

But in the midst of all of this randomness, there is one absolute certainty: In the long run, the casino itself *always* makes money. Casinos are truly cash cows. When people debate the ethics of government-sponsored betting, concerns are often raised about the social impact of gambling houses, the problem of gambling addiction, and the message sent to our children. But no one ever raises the concern that the casino will lose money; that simply never happens.

How can this be? How can such certainty arise from so much uncertainty? With individual players' fortunes being so random, so unpredictable, how is it that the casino itself profits at such a steady, predictable rate?

The answer to this question involves two key facts. The first is that the casino's overall performance is the combined result of many, many individual bets. Each hour, there may be hundreds or even thousands of bets on roulette, and thousands more on the slot machines and the other games. Whereas any individual gambler might make just a few bets here and there, the casino as a whole is involved with a huge number of separate bets.

When random events are repeated over and over again, the fraction of successes gets closer and closer to their average value (also called *expected value*). Thus, if you repeatedly flip a fair coin, in the long run about half of the flips will be heads. Or if you repeatedly toss a fair six-sided die, in the long run about one-sixth of the rolls will come up 5. This isn't just conjecture, it's the law: the *Law of Large Numbers*. This law says that if you repeat any random event often enough, over the long run the good luck will cancel out the bad, and you will be left with approximately the "correct" average; that is, an average that mimics the true probabilities.

In Tom Stoppard's play *Rosencrantz and Guildenstern Are Dead,* a coin is repeatedly flipped, hundreds of times, over the course of several days. Every single time, the coin comes up heads. This is one of many occurrences that disturb the central characters' sense of reality. Are they right to be concerned? Absolutely! The Law of Large Numbers guarantees that, over the long run, good and bad luck tend to cancel each other out, and averages tend to settle down to their true probabilities. Thus, when you flip a coin many, many times, you will get approximately half heads and half tails. Events like hundreds of heads in a row simply never occur.

Of course, the coin doesn't know that the tails should cancel out the heads. Indeed, even if you have just gotten heads four times in a row, the coin is still equally likely to show heads or tails on the next flip. What the Law of Large Numbers says is that, even though each coin flip is equally likely to be heads or tails, without regard to what came before, nevertheless the fraction of heads will get closer and closer to one half, in the long run.

The Law of Large Numbers says that if a gambling game is even very slightly to your advantage *on average,* and you play it long enough, you are sure to come out ahead. On the other hand, if a game is even very slightly to your disadvantage *on average,* and you play it long enough, you are sure to come out behind. Even though each individual game is independent, with no regard for what came before, nevertheless in the long run, all that matters is the game's *average* amount won or lost.

The second key fact is that in every professional casino, every single betting game is weighted ever so slightly in the house's favor. Each gambling game's payout rules are designed specifically so that, even though on any one bet anything can happen, over the long run, on average, the house will come out slightly ahead.

If a casino has only a few customers, and they each make only a few bets—perhaps even large bets—there is no telling what will happen. The customers might lose their bets or they might win. The casino might make money or it might lose money. The outcome is random, with no guarantees.

On the other hand, if a casino has many customers making many different bets, randomness virtually disappears. The Law of Large Numbers says that because each individual bet is weighted slightly in the house's favor, it is a certainty that the house will make money over the long run.

In short, to make a profit, the casino doesn't have to be lucky, just patient. While the players might base their gambling hopes on the illusion of having a "hot hand" or a "lucky number," or on the alignment of the planets, the casino can afford to base its hopes on something more certain: the Law of Large Numbers.

The Honorable Bus Fine

You're on a European bus where payment is handled by a sort of honor system. No one collects fares when riders board, but they are expected to buy a ticket anyway. Occasionally, inspectors arrive to check tickets and they give a fine to any ticketless riders.

Being an honest person, you have bought your ticket. However, while waiting for the bus, you see Cheapskate Charlie loitering by the ticket machine, so close that you can overhear his mutterings.

"Hmmm, it costs one euro for a bus ticket. If I don't have a ticket, and I get caught, then I will get a 10-euro fine. However, there's only about one chance in 20 that I will get caught. So, in the long run, I'll only have to pay the 10 euros about one time in 20. That works out to just half a euro per ride. So, according to the Law of Large Numbers, in the long run it's cheaper for me to never buy a ticket."

Satisfied and smug, Charlie enters the bus with no ticket. You're not looking for a fight, so you say nothing. However, that night you write a letter to the local transit authority, suggesting that they increase their fine to at least 20 euros immediately.

Calculating the Averages: Roulette and Keno

For a casino to profit from the Law of Large Numbers, it has to make sure that every single betting game is weighted slightly in its favor. One slip-up, one game that is weighted just barely in the player's favor, and the casino might lose millions of dollars in the long run.

How does the casino guarantee that the games are weighted in its favor? It employs experts in probability theory to calculate the average, or expected, net payout for each game.

To see how this works, consider the simplest example: roulette. A standard (American-style) roulette wheel consists of 38 spots: the numbers 1 to 36 (alternately red and black), plus the special spots 0 and 00 (which are green). The wheel is designed so that when it is spun, the marker ball is equally likely to land in any one of the 38 spots.

A player places a bet on where the ball will drop on the next spin. There are a variety of different betting options, each having a different payout. For example, a player might choose to bet $10 on red, meaning that he will win $10 if the ball lands on one of the 18 red spots, but will lose $10 if the ball instead lands on a black or green spot.

What happens on average? Well, there are 18 red spots out of 38 total spots, so 18 times out of 38, the player will win $10. However, there are 20 black or green spots out of 38 total spots, so 20 times out of 38, the player will lose \$10. Therefore, the player's average winning on the bet equals \$10 × 18 ÷ 38 minus \$10 × 20 ÷ 38. This works out to −\$0.526; that is, on average, the player will lose just over 52 cents by making this bet.

Of course, the player will never actually lose 52 cents. On each bet, he will either win \$10 or lose \$10. However, if he makes lots and lots of such bets, over and over again, in the long run, on average, he will lose about 52 cents for each bet he makes. The Law of Large Numbers claims another victim. These same odds apply to all of the casino's thousands of customers, ensuring huge profits for the casino.

Another way to think about long-run gambling is in terms of the well-known *gambler's ruin* question. Suppose you start with \$1,000, and repeatedly make \$10 bets on red. What is the probability that you will double your holdings—that is, win another \$1,000—before you lose the \$1,000 you started with? Since you have almost a 50% probability of winning each roulette bet, you might expect to have almost a 50% probability of winning \$1,000 before losing \$1,000. In fact, the probability of winning \$1,000 before losing \$1,000 when repeatedly making \$10 roulette bets is just one chance in 37,650—a tiny probability. So, if you keep betting on roulette, it is virtually impossible that you will make \$1,000 before losing your initial \$1,000 capital.

What about the other possible roulette bets? Certain other bets—such as betting on black, or on the odd numbers, or the even numbers, or the numbers 1 to 18, or the numbers 19 to 36—work exactly the same as the bet on red. In each case, there are 18 spots on which you win and 20 spots on which you lose, so on average you will lose just over 52 cents for every \$10 bet.

Of course, there are many other kinds of roulette bets. For example, there are the "dozen" bets, where you bet on a collection of 12 numbers. If you bet \$10 on the "second 12" numbers (13 to 24) and the ball lands on one of the spots 13 to 24, you win \$20. If the ball lands on one of the other spots, you lose just your \$10 bet.

Wow, you might think, this is great. If I sometimes win \$20 and sometimes lose just \$10, surely the odds are in my favor.

Unfortunately, no. Because there are just 12 spots in the range 13 to 24, just 12 times out of 38 will you win \$20. The other 26 times out of 38, you will lose \$10. Hence, your average winning on this bet is equal to $\$20 \times 12 \div 38$ minus $\$10 \times 26 \div 38$. This again works out to $-\$0.526$, the same as before. So the odds are still against you on this bet. And, in the long run, you will again lose about 52 cents for every bet that you make, just like when you bet on red.

What about betting on just a single number? If you bet \$10 on, say, 22, then you win \$350 if the ball lands on 22. Otherwise, you just lose your \$10. This is it, you figure. If you bet for a while, then eventually you will win a \$350 payout, at which point you will surely come out ahead.

Maybe not. There is just one chance in 38 that the ball will land on 22, while there are 37 chances that the ball will land on some other spot. Therefore, if you bet \$10 on 22, then your average payout is equal to $\$350 \times 1 \div 38$ minus $\$10 \times 37 \div 38$. This works out to (you guessed it) $-\$0.526$. Once again, if you repeatedly make this bet, in the long run you will lose about 52 cents for every bet that you make.

Table 3.1 Probabilities and Payouts for \$10 Roulette Bet

Bet	Probability	Payout	Expected Loss
Red/Black/Even/Odd	18/38	\$10	\$0.526
Dozen	12/38	\$20	\$0.526
Single	1/38	\$350	\$0.526

Darn, those roulette-wheel designers are pretty clever. They have designed every game so that, overall, the odds are in the casino's favor. They're not *too* much in the casino's favor, or no one would ever bother to bet. But they are just enough in the casino's favor to guarantee a long-run profit margin, typically between 1% and 3% of the total amount of money bet. And their success is all due to the Law of Large Numbers. With such a huge amount being bet over the course of a year, this small profit margin really adds up: worldwide annual gambling revenues amount to hundreds of billions of dollars.

Of course, you might get lucky. The ball might land on 22 on the first spin. Or perhaps on the second or third spin. If so, then hurry up and cash in your chips, and leave! But don't count on this occurring. Over the long run, casino gamblers will lose more money than they will win. Some gamblers will come out ahead, but on average the casino will make money. It has to be that way. It's the law.

Slowly but Surely?

In Aesop's fable "The Tortoise and the Hare," those two animals have a race. Although the hare is much faster, he is also careless: part way through the race, he decides (given his substantial lead) to have a quick nap. While he is sleeping, the tortoise rumbles by, slow and steady, and chugs ahead to victory.

It is customary, when recounting this tale, to blame the hare for being too inconsistent, too unpredictable, too random. By contrast, the slow-and-steady tortoise is portrayed as a model of hard work and discipline, proving that we can all succeed in life if only we focus steadily and seriously on our task.

The Probability Perspective provides a more precise understanding. The Law of Large Numbers tells us that the real issue here isn't which of the tortoise or hare is more reliable, or more consistent, or a better role model. The issue is which of them goes faster *on average*. Whoever goes faster on average is guaranteed to win races in the long run.

Suppose that the tortoise always goes along at a steady pace of 1 kilometer per hour. The hare, on the other hand, can run 4 kilometers per hour when he isn't napping. Who will win the race?

If the hare is very lazy and naps for four out of every five hours (on average), he is in trouble. In this case, in every five-hour period the hare will only run for one hour and thus only cover 4 kilometers. By contrast, the slow-and-steady tortoise will progress 5 kilometers every five hours. Victory for the tortoise!

On the other hand, if the hare naps in just half of the available hours (on average), then in every two-hour period he will still run for one hour and advance 4 kilometers. Meanwhile, the tortoise will advance just 2 kilometers. Victory for the hare!

So, from a Probability Perspective, the story of the tortoise and the hare is about balancing fast speeds and nap times, by carefully computing the average values to determine who will win the race. It isn't about a steady work ethic or the evils of an unreliable lifestyle. Such an interpretation amounts to blatant, hurtful, unfair discrimination—against randomness itself.

Another casino game is keno. In the 80-ball version, a player might select 10 of the numbers between 1 and 80. Twenty balls numbered between 1 and 80 are then chosen at random by a "blower" machine, and the player's payout depends on how many of those 20 numbers she selected, or her number of matches. In one version of the game, if you bet $10, you receive nothing for three or fewer matches, but you receive $10 for four matches, $20 for five matches, $200 for six matches, $1,050 for seven matches, $5,000 for eight matches, $50,000 for nine matches, and a whopping $120,000 for matching all 10 of your choices.

A potential payout of $120,000 for a mere $10 bet sounds very tempting, and helps to draw in customers. Unfortunately, the number of different possible choices of 20 balls out of 80 that the blower can make is nearly 4 followed by 18 zeros, a huge number. And each choice is equally likely to occur. The number of these choices that include all 10 of your selections is nearly 4 followed by 11 zeros—still a huge number, but much smaller. The probability of matching all 10 numbers in keno is the second number divided by the first, which works out to about one chance in 9 million—virtually impossible. I'm sorry to say that you just won't match all 10 numbers anytime soon.

Now, matching four numbers is more likely. The number of choices that match just four of your selected numbers is equal to 521 million billion, and dividing that number by the total number of choices works out to a probability of nearly 15%. On the other hand, matching four numbers just gets you your $10 bet back, nothing more. The probability of getting five matches, and thus making a $10 profit, is about 5%; and for six matches the probability is just over 1%.

Table 3.2 indicates the expected return (the payout multiplied by your probability of winning it) from all the different numbers of matches. Adding them all up tells us the overall expected payout from betting $10 on keno, namely $7.49. This means that, on average, for every $10 that you bet at keno, you will get back about $7.49. Or, to put it differently, on average you will lose $2.51. Once again, the odds are stacked against you.

Table 3.2 Probabilities and Payouts for a $10 Keno Bet

# Matches	Probability	Payout	Expected Return
0	4.58%	$0	$0
1	17.96%	$0	$0
2	29.53%	$0	$0
3	26.74%	$0	$0
4	14.73%	$10	$1.47
5	5.14%	$20	$1.03
6	1.15%	$200	$2.30
7	0.16%	$1,050	$1.69
8	0.014%	$5,000	$0.68
9	0.00061%	$50,000	$0.31
10	0.000011%	$120,000	$0.01
Total:	100%		$7.49

More money is bet on slot machines than on any other casino game; some 60% of casino profits come from the slots. I find this surprising, since slot machines operate on hidden mechanisms. There are no spinning wheels, bouncing balls, or rolling dice, and no way to directly check the probabilities. Whether it's a traditional machine of gears and levers or a modern, computer-controlled video lottery terminal, playing a slot machine requires a certain level of trust that the casino operator isn't cheating.

Manufacturers of slot machines do make claims about the probabilities of payouts. The probabilities vary from one type of machine to the next, but typically they claim that the customer will recover between 88% and 95% of their bet, on average. Or, to put it differently, for every $10 you bet in slot machines, you can expect to get between $8.80 and $9.50 back, and to lose between $0.50 and $1.20.

Rolling the Dice

Many games involve rolling dice. While no one can say for sure what the dice will show, we can still improve our chances by considering probabilities. If you roll one ordinary six-sided die, it is equally likely to show any of the numbers 1, 2, 3, 4, 5, 6. But in many games, two dice are rolled, and the two numbers are added together. The result could be anything from 2 (if you roll a pair of 1's) to 12 (if you roll a pair of 6's). But are all possibilities equally likely?

No, they are not. When rolling two dice, there are a total of 36 possibilities for the pair of numbers that you get, on die #1 and die #2 respectively, as shown in Table 3.3.

Table 3.3 Possible Number Pairs for Two Dice

(1,1)	(1,2)	(1,3)	(1,4)	(1,5)	(1,6)
(2,1)	(2,2)	(2,3)	(2,4)	(2,5)	(2,6)
(3,1)	(3,2)	(3,3)	(3,4)	(3,5)	(3,6)
(4,1)	(4,2)	(4,3)	(4,4)	(4,5)	(4,6)
(5,1)	(5,2)	(5,3)	(5,4)	(5,5)	(5,6)
(6,1)	(6,2)	(6,3)	(6,4)	(6,5)	(6,6)

These 36 pairs are all equally likely. A quick count shows that just one of them, (1,1), has a sum of 2. This means that the probability of obtaining a sum of 2 is just one chance in 36. On the other hand, six of them have a sum of 7: (1,6), (2,5), (3,4), (4,3), (5,2), and (6,1). So, the probability of a 7 is 6/36. Table 3.4 shows the probabilities for every sum that can be rolled with a pair of dice.

Table 3.4 Probabilities When Rolling Two Dice

Sum	# Pairs	Probability	Percentage
2	1	1/36	2.78%
3	2	2/36	5.56%
4	3	3/36	8.33%
5	4	4/36	11.1%
6	5	5/36	13.9%
7	6	6/36	16.7%
8	5	5/36	13.9%
9	4	4/36	11.1%
10	3	3/36	8.33%
11	2	2/36	5.56%
12	1	1/36	2.78%
Total:	36	36/36	100%

So, if you roll two dice, the most likely value of the sum is 7, which will occur one time in six, or about 17% of the time. Next most likely are 6 and 8, each of which will occur five times in 36, or about 14% of the time. Five and 9 will each occur about 11% of the time, with other numbers less likely. At the extremes, 2 and 12 will each occur just one time in 36, or less than 3% of the time.

Here is another way of seeing why 7 is the most likely result. No matter what number the first die shows, the second die *always* has some choice that will make the sum be 7. If the first die is 1, then the second die could be 6. If the first die is 2, then the second die could be 5. And so on. So, no matter what the first die shows, there is always one chance in six for the sum to equal 7. All other sums are less likely. For example, if the first die is 1, it is impossible for the sum to equal 8. Or, if the first die is 6, it is impossible for the sum to equal 6. That is why 7 is the most likely outcome.

A more intuitive way to see why 7 is the most likely result is to use, once again, the Law of Large Numbers. The average value when rolling one die is the average of all of the numbers from 1 to 6. This is equal to the number exactly halfway between 1 and 6, which is 3.5 (not 3, as some people think). Thus, the average when rolling two dice is twice 3.5, or 7. The Law of Large Numbers tells us that the result is most likely to be close to its average, or close to 7 in the case of two dice. If you roll more dice, the result will be even more likely to be near its average. If you roll 10 dice, the result could be anything between 10 and 60; however, the most likely sum is 35, with sums close to 35 much more likely than sums that are much higher or much lower.

This basic understanding of dice can help us to make better decisions in games of chance and increase our probability of winning. For example, suppose you are playing Monopoly and decide that now is the time to build hotels for one of your monopolies. Then when opponents land on your hotel, they will have to pay you lots of money.

Suppose you own two monopolies, one yellow and one orange, each consisting of three properties. One opponent is within striking range of the yellow monopoly, and he will land there if he rolls a 2, 3, or 5 on his next turn. Another opponent is approaching the orange monopoly, and she will land there if she rolls a 6, 8, or 9. On which monopoly should you build?

The numbers 6, 8, and 9 are all close to 7, which is the most likely roll. By contrast, the numbers 2, 3, and 5 are on average much farther away, and therefore much less likely. The orange monopoly is considerably more likely to get visited on the next turn, so the smart move is to build there. This probability won't guarantee you success every single time, but it will help you win more games over the long run.

Similar reasoning applies to virtually every game involving dice. In the recently popular game Settlers of Catan, the playing area consists of regions with numbers printed on them. Resources are distributed by rolling two dice and giving resource cards to those players who are adjacent to areas with the corresponding number. Good Catan players always crowd around the areas with numbers like 6, 7, and 8. They know these numbers are the most probable outcomes of the dice rolls, and thus the regions likely to provide the most resources over the long run.

Suppose now that you repeatedly roll dice, and want a certain number to appear at least once. What are the odds? For example, if you roll a die once, the probability of getting a 3 (or any other number between 1 and 6) is one chance in six. But suppose you roll a die four times. What is the probability of getting at least one 3 in the four rolls?

Many people think the answer must be 4/6, or 67%. They reason that the probability of getting a 3 after four rolls must be four times as great as the probability after just one roll. But this can't be. By that logic, if you rolled a die six times, the probability of getting at least one 3 would be 6/6, or 100%—and we all know *that* isn't true (even after six rolls, you are not *guaranteed* to get a 3). The problem is that this reasoning involves overcounting; if you get 3 on four different rolls, you are still effectively adding in a 1/6 probability for *each* of the four 3's that you got, even though you should really just count it once.

Here is the correct way to calculate the probability of getting at least one 3 when rolling a die four times. Each time, the probability of *not* getting a 3 is equal to 5/6. So, on four rolls, the probability of not getting a 3 all four times is equal to four copies of 5/6 all multiplied together, which equals 48.2%. The probability of getting at least one 3 is equal to 100% minus 48.2%, or 51.8%. This is just *barely* more than 50%, and is far less than the 67% that many people think the answer should be.

Remarkably, this very problem is what began the modern mathematical theory of probability. In seventeenth-century France, a shrewd gambler named Antoine Gombaud, Chevalier de Méré, was making a handsome profit by betting people that at least one 6 would come up if a die was rolled four times. (We now know that his probability of victory was 51.8%, which is more than 50%, so his long-run profit was guaranteed by the Law of Large Numbers.) He then tried modifying the bet to say that a pair of 6's would appear at least once, if a pair of dice was rolled 24 times. He reasoned that since each pair of 6's has probability 1/36, and since 24/36 is equal to 4/6, the probabilities of the two games would be the same, and his winning ways would continue. However, the true probability of winning this second bet is equal to 100% minus 24 copies of 35/36 all multiplied together, which works out to 49.1%. This is a little bit less than 50%, and the poor chevalier started losing money. He became a victim of the very Law of Large Numbers from which he had earlier profited.

Puzzled, de Méré contacted French mathematician and philosopher Blaise Pascal. Pascal in turn corresponded about this and related questions with the brilliant mathematician Pierre de Fermat, a Toulouse government official and lawyer who later on would propose the infamous *Fermat's last theorem*. Their letters are now regarded as the first serious attempt to study probability and uncertainty in mathematical terms.

The Curious Case of Craps

There is another casino game that is curious both for its complicated rules and for its interesting probabilities: the game of craps. Craps is played by repeatedly rolling a pair of ordinary six-sided dice and considering the sum each time. If the sum is equal to 2, 3, or 12, the player loses immediately. If the sum is equal to 7 or 11, the player wins immediately.

If the sum is any other value (say, 4), that value becomes the player's "point." The player then repeatedly rolls the pair of dice until such time as the sum either equals the point—in which case the player wins—or equals 7, in which case the player loses.

To summarize, on the first roll, 2, 3, and 12 are bad, while 7 and 11 are good. Any other sum sets off a race to repeat that sum before rolling a 7. If, for example, the player has bet $10 and wins, he receives another $10; if he loses, he forfeits the $10 bet.

Most people's reaction upon first hearing these rules is that craps is a strange, complicated game. Where did these rules come from? Craps originated in a French variant of an older English dice-rolling game, and it was gradually refined to its present form on American riverboats and in gambling houses. (It is believed that the name *craps* was a mispronunciation by the French of the English word *crabs,* which was slang for rolling a pair of 1's.) But why these particular rules?

To answer this question, we have to consider—you guessed it—probabilities. From the Law of Large Numbers, we already know the basic principle of casinos: every game should be weighted in the casino's favor, to guarantee long-run profits. But it shouldn't be *too* heavily weighted, or no one will play. So, how does the game of craps stack up?

From Table 3.4, we know that the probability of rolling a 7 or 11 on the first roll (and thus winning immediately) is equal to eight chances in 36, or about 22.2%. On the other hand, the probability of rolling a 2, 3, or 12 on the first roll (and thus losing immediately) is equal to four chances in 36, or about 11.1%. Thus, on the first roll, you are twice as likely to win as to lose.

So far so good. But what about the other 66.7% of the time, when you roll some other value on the first roll? Here things get complicated, because the value on the first roll becomes your "point," and the probability of winning depends on what your point equals. Calculating the probability of winning now requires computing a complicated sum of different probabilities of obtaining the various point values, multiplied by the probabilities of winning once you know your point.

When all the fractions have been multiplied and sums have been computed, it turns out that the overall probability of winning at craps is equal to 244 chances in 495, or 49.2929%. In other words, the probability of winning at craps is just *barely* less than a fair 50–50 bet.

This means that if you bet $10 on craps, then 49.2929% of the time you will win $10, while the other 50.7071% of the time you will lose $10. Your average winning works out to $10 × 49.2929% minus $10 × 50.7071%. This is equal to −$0.141. Thus, if you play craps over and over again, in the long run you will lose about 14 cents for every $10 bet that you make. The odds are just barely weighted against you, but the Law of Large Numbers says that is still good enough that in the long run the casino will win—and all of its gamblers, taken together, will lose—a lot of money.

We see now how the rules for craps developed. If the odds were weighted toward the player, the casinos would lose money and eventually have to change the rules. But if the odds were weighted too heavily toward the casino, players would get frustrated and stop playing, and the casino would also eventually have to change the rules. Finally, with enough tweaking, the rules were fixed so that the player's probability of winning was just under 50%, thus guaranteeing good fun for the player and steady profits for the casino.

The game of craps has another interesting aspect: spectators can get involved, too. Other gamblers, while watching the main player roll the dice, can make various kinds of "side bets" on the game. The results of the side bets depend on what the player rolls. (That is why one often hears a big cheer from a craps table in a casino. The spectators are not expressing any great affection for the player; rather, they are cheering because their own side bet has paid off.)

One kind of side bet is particularly intriguing (at least to probabilists). A gambler can bet that the player throwing the dice will lose. Officially, this is referred to as betting on the "Don't Pass Line." The rule is, if you bet $10 on the Don't Pass Line, and the player loses at craps, you receive $10. If the player wins, you forfeit your $10 bet.

Betting that a player will lose may seem rude or even hostile, but it seems to provide an opportunity for patrons to cash in on the casino's guaranteed profits. If you bet against a player, you are effectively adopting the casino's position in the game. And since the casino is guaranteed to win money in the long run, won't you benefit from this same guarantee?

Don't drop this book and rush out to the nearest casino just yet. As you might suspect, there is a catch. There is one little, subtle additional rule for betting on the Don't Pass Line. If the player rolls a 12 on the very first roll, the player loses (of course). However, in that one special case, even though the player loses, you don't win your Don't Pass Line bet. Instead, your bet is considered a tie: your $10 is refunded but you don't receive another $10.

Big deal, you might think. After all, there is only one chance in 36 that a player will roll a 12 on the first bet. And, even if he does, you don't actually lose, you just get a refund and get to bet again. What is the harm in that?

Unfortunately, this one special rule is just enough to change the Don't Pass Line bet from being weighted in your favor to being weighted in the casino's favor. This rule change means that one time in 36 you will miss out on $10 that you would otherwise have won. So, this reduces your average winnings by $\$10 \times 1 \div 36$, or $0.278.

Now, 27.8 cents isn't much. But remember that when playing craps, the player will lose 14.1 cents on average and the casino will win 14.1 cents on average. So, with a $10 bet on the Don't Pass Line, on average you will win 14.1 cents minus the 27.8 cents for the rule change. Overall, on average you will lose 13.7 cents.

As a result, even the Don't Pass Line bet is weighted against you. That one little rule change was just enough to ensure that, whether you bet to win or bet on the Don't Pass Line, on average you will lose money. No matter the game, the odds are always against you.

Table 3.5 Average Loss When Betting $10 at Various Casino Games

Game	Average Loss
Roulette	$0.526
Keno*	$2.51
Slot Machines	$0.50 to $1.20
Craps	$0.141
Don't Pass Line	$0.137

*Specifically, the version of keno described above; others may vary.

Of course, various other side bets are allowed in craps. They have exotic names like Don't Come bets, Proposition bets, and Hardways. Some of them are complicated, and computing their probabilities isn't always easy.

But even without computing any probabilities, you can rest assured that with each of these bets, on average the gambler will lose money. After all, the casinos aren't stupid: they hire consultants to make sure that each and every bet is weighted slightly in their favor. This is what guarantees them steady profits in the long run.

Betting Temptations

"Step right up," the barker barks. "Make your fortune using the wonders of gambling. What do you say to a little roulette, mister? Why, just last week one roulette player made a $10,000 profit!"

Politely you decline, explaining that in roulette the odds are stacked against the customer in the long run.

"How about keno, then? Or craps?" You explain that you know the odds are against the customer in those games, as well.

"Okay, then," the barker continues. "I see you're too smart for simple games like those. Why don't you play our brand-new game, Spin–Roll–Flip–Blow? It combines a roulette wheel, dice, coins, and a blower machine in such a complicated way that you could never figure out the odds. What do you say to that, then?"

You look around the casino to see the glittering gold trim, the plush carpet, the free drinks, and the hundreds of well-paid employees. You know that the owners are making a steady profit. This must mean that the odds in all of their games are stacked against the customer, at least a little.

"I don't think so," you reply, exiting the casino to go swing dancing instead.

A Life of Large Numbers

Once we understand the Law of Large Numbers—the principle that in the long run randomness cancels out—we notice this law appearing in many different aspects of our lives.

For example, most of us like to get a second opinion (whether from doctors or from friends), read more than one newspaper, and get advice from more than one stockbroker. Why? The answer is simple: any one doctor (or friend or newspaper or broker) might be wrong or biased or stupid or eccentric or just be having a bad day. The more ideas or events or outcomes we average together, the more such randomness tends to cancel out, and the more certain we can be about the conclusions we reach. Thus, every time you seek a second opinion, you are employing a version of the Law of Large Numbers.

The Law of Large Numbers similarly explains why, if you play Scrabble once, anyone might win; but if you play over and over again the best player will come out ahead. If you flip a few coins you might get any result, but if you flip lots of coins you will get about half heads and half tails. If you invest in one stock, you may win or lose a tremendous amount, but if you invest in many stocks, your fortunes will follow the overall market trend. At any one intersection you might get stuck at a red light, but over time traffic lights treat every driver equally. If you sprinkle just one grain of salt on your pasta it might land anywhere, but if you keep sprinkling randomly the salt grains will spread out nicely on your plate. And each time you take a breath, you can rest assured that of the trillion trillion or so molecules of oxygen in the room with you, plenty of them are close enough to breathe, and not all of them are hiding under the bed. In every case, the randomness of each individual game or coin flip or stock fluctuation tends to cancel out as we take bigger and bigger averages.

It is difficult to escape from the Law of Large Numbers, even if you try. On a recent vacation my wife and I stayed at a winter resort, and I tried to unwind from my life as a probabilist. While relaxing in a steaming outdoor hot tub surrounded by snow, I noticed a chemical dispenser floating lazily in the tub, dispensing disinfectant and who knows what else. At any one moment, the dispenser could be anywhere. But over time, with enough random pushes and pulls and disturbances, the dispenser would travel all around the tub, and the Law of Large Numbers would ensure that the tub is equally disinfected on all sides. Even on vacation, I found probability theory floating beside me.

The principle of long-run averaging also applies in many other areas. We shall see that it explains why small differences in poker strategy can have a big effect over a long night of playing. It explains why medical studies can sometimes demonstrate the superiority of one treatment over another. It also explains how opinion pollsters can conclude that "these results are considered accurate within four percentage points, 19 times out of 20."

So, does the Law of Large Numbers solve every problem in probability? No, not quite. Indeed, the celebrated economist John Maynard Keynes had this to say about relying too much on long-run calculations: "This long run is a misleading guide to current affairs. In the long run we are all dead. Economists set themselves too easy, too useless a task if in tempestuous seasons they can only tell us that when the storm is long past the ocean is flat again."

Keynes has a point. Just as it is not sufficient for athletes "usually" to perform well—they have to perform well specifically at the Super Bowl or the Olympics, where it counts the most—it is also not sufficient to know that the economy will improve "in the long run"; what will happen over the short term is also important. But despite this limitation, there is still enormous benefit and power in long-run probability calculations. From casinos to evolution, from polling to poker, the Law of Large Numbers provides a deep understanding of the long-term effects of randomness.

4

Dealing the Cards

Bridge, Poker, and Blackjack

It seems that many people just can't get enough randomness. Even though our lives are already filled with uncertainty, we love to pile it on by playing games of chance. Cards, dice, and spinning wheels are used to inject uncertainty into competitions, to make them more exciting and fun. But what if you also enjoy winning games? How can you apply skill and logic to the randomness of a game of chance?

The Law of Large Numbers provides a partial answer. It tells us that over the long run, the player who will win the most is the one who has the highest probability of winning each game. So, when playing games of chance, your goal should always be to make decisions and adopt strategies that will increase the probability that you will win. You might not win every game, but over the long run you will get your just rewards.

Even non-random games are often won because of probabilities. For example, Boston Celtics basketball great Larry Bird was famous for, among other things, his excellent free-throw shooting. Now, Bird sometimes scored his free throws and sometimes missed. Meanwhile, I, playing on low-level student intramural basketball teams, sometimes scored my free throws and sometimes missed. So what was the difference between Bird and me? Probabilities! Bird made 88.6% of his free throws over his career, including an astounding 71 in a row during the 1989–90 NBA season. I, on the other hand, didn't score more than 50% of my own free throws (actually it was probably even less, but let's not dwell on that). So both Bird and I shot free throws in the 1980s, and for both of us the outcome was random. However, Bird had a much higher probability of success than I did.

Or, consider bowling. Championship bowlers get strikes about two thirds of the time, whereas mediocre bowlers like me get strikes less than one third of the time. There are no other game aspects like passing or defense or head-to-head competition; each player bowls separately. So in bowling, as with basketball free throws, the difference over the long run between winners and losers is simply a question of probabilities.

Building a Bridge

Bridge is a complicated card game with many different aspects, including sophisticated bidding systems that take years to master. However, bridge also uses a tremendous amount of probability. Good players know the odds of success of various alternatives, and they are always choosing the most promising options, to increase their probability of winning.

When playing bridge, the bidding determines which player becomes the "declarer." This player gets to see her own 13 cards (of course), plus all 13 cards of her partner. However, the 26 remaining cards are not visible, and are divided among the two opponents in a random way. The declarer must decide which cards to play when, without knowing which opponent has which cards.

A typical situation might involve the declarer wondering which opponent has the King of Spades (Kings rank second only to Aces in bridge). If she guesses correctly, then she can win an extra trick using a bridge maneuver called a finesse (by which she lures the correct opponent into playing his King and tops it with her Ace), and succeed in making her contract. But if she guesses wrong, she will lose the trick and the contract. What should she do?

A novice player might just guess at random, and have a 50% chance of success. However, an expert will be more careful, looking for clues about where the missing King of Spades lies. If one opponent has previously bid and indicated that he has lots of high cards, that opponent is more likely to have the King. On the other hand, if one opponent is seen to have lots of Hearts, that suggests he has fewer Spades, so is less likely to have the King. A clever player, looking carefully for clues, might be able to increase her chance of success from 50% to 60%, or 70%, or even higher. Such improvements have only a small effect on any one hand of bridge, but over the long run, they separate the strong players from the weak.

While a good bridge player can improve the probabilities, she still has a lot of randomness to overcome. The total number of different ways in which a deck of 52 cards can be divided into four hands of 13 cards each is unimaginably huge—larger than the number written as a 1 followed by 28 zeros. Some of those deals will be favorable to the player, and some will be unfavorable. Over the long run, by the Law of Large Numbers, the better players will win more often. However, even with the greatest of skill, it might still take a long time for the randomness of the different deals to cancel out.

To overcome this problem, serious players often play a version of bridge called *duplicate bridge*. The cards are carefully managed (by a tournament director) so that all players who are playing the same position (North, South, East, or West) play the exact same hands over the course of a session. So, if in one round you, as North, get the Ace and Queen of Spades, and the Ten of Diamonds, and so on, then in another round I, also as North, would play with those exact same cards, against another pair of opponents. When you and I later compare our results, we can see who performed better with exactly the same cards. Much of the randomness is eliminated, and the players' skill more easily shines through. Serious bridge players prefer this version; their bridge abilities are measured more precisely because the additional "luck factor" of the quality of the cards they are dealt is eliminated. But novice players often find duplicate bridge more stressful because they cannot blame bad cards when they lose.

So is duplicate bridge completely free of luck? No. Even if different players use the same cards, some playing strategies will happen to work out better than others, just by luck, since the opponents' cards remain unknown. However, in duplicate bridge randomness is greatly reduced. In either conventional or duplicate bridge, the Law of Large Numbers guarantees that the best players will win the most often in the long run. But with duplicate bridge, the Law of Large Numbers kicks in a lot faster, and the better players rise to the top more quickly.

Bridge Bickering (A True Story)

I once commented to a fairly serious bridge player that although duplicate bridge is mostly based on skill, it still involves some luck. He vigorously disputed this. "There is no luck in duplicate bridge!" he exclaimed.

"Oh, come now," I pressed on. "Suppose a certain finesse has just a 35% chance of success, so you do not try it. But some other, weaker player tries it anyway, and he happens to succeed."

The serious player scoffed. "If someone succeeds with a 35% finesse, then that shows nothing. It's completely meaningless. He just got lucky!"

I smiled and paused, while the serious player slowly realized that he had just proved my point.

Poker Power

No game of chance has captured the popular imagination quite like poker. Poker showdowns feature in numerous movies about the Wild West, from *Butch Cassidy and the Sundance Kid* to *Maverick*. Psychological manipulation, macho posturing, tough talk, and the occasional six-shooter all combine to give poker its entertainment value. Such depictions of poker emphasize the competitive, psychological, and financial aspects of the game, all of which are significant. Unfortunately, they leave out the most important part: probabilities.

In many a movie scene, the final hand is won by the hero when he is dealt a Royal Flush: the 10, Jack, Queen, King, and Ace all of the same suit. Is this really a plausible outcome? In reality, there are nearly 2.6 million different five-card hands that can be dealt. Of these, just four are Royal Flushes (one for each suit). So, the probability of the hero's actually being dealt a Royal Flush on the final hand is four chances in 2.6 million, or about one chance in 650,000. It is extremely rare. And, popular entertainment aside, no amount of tough talk or intimidation (short of outright cheating) will change this probability. From the strongest cowboy to the cleverest card shark to the greenest novice, we all have exactly the same chance of being dealt a Royal Flush in five cards.

Of course, there are many different versions of poker, including some with wild cards, extra cards to choose from, opportunities to exchange cards, and so on. Each of these modifications changes all the probabilities and, with enough wild cards, even a Royal Flush isn't so unlikely. Nevertheless, the probabilities are the same for everyone. And the way to succeed at poker is not by magically getting a Royal Flush every time, but rather by understanding probabilities and making good decisions once the cards are dealt.

Suppose you're playing five-card stud poker, where each player is dealt a total of five cards (with no extra or wild cards). Suppose you have been dealt four Spades, with the fifth card still to come. If the fifth card ends up being a Spade, then you will have a Flush (all five cards of the same suit), which is a very good hand that will likely win the pot. On the other hand, if the fifth card is not a Spade, then your hand will be very weak (at best, one pair) and will probably lose. Everything comes down to whether or not your fifth card will be a Spade.

What is your probability of success? If the deck is well shuffled and nobody cheats, every unseen card is equally likely to be dealt next. You have already seen four cards, all Spades. So, there are 48 cards remaining, of which nine are Spades. This means that the probability that the next card will be a Spade is equal to 9/48, or about 19%. Those odds are pretty low, so perhaps you should fold at this point (though that decision also depends on the pot odds, discussed below).

Is that all there is to it? No. In many poker games, some of the cards are dealt face up, for everyone to see. For example, in the usual version of five-card stud, all but the first of each player's cards are viewable by all. Consider again the Flush example, where you have been dealt four Spades and are waiting for a fifth. Suppose you are at a table with nine opponents, each of whom already has three face-up cards showing. That makes an extra 27 cards that you already know about, leaving just 21 unseen. If none of your opponents' face-up cards are Spades, then there are still nine unseen Spades. So, in this case, the probability that your fifth card will be a Spade increases to 9/21, or 43%—far more than the 19% it was before. By contrast, if seven of those other 27 cards are Spades, then there are only two Spades left, so your probability reduces to 2/21, or 9.5%, which is far worse.

For another example, suppose you have been dealt a 5, a 6, an 8, and a 9. If your next card is a 7, then you will have a Straight (five cards in succession), a good hand that will probably win. What is the probability that your fifth card will be a 7? Well, there are only four 7's in the deck. If you haven't seen any other cards, then your chances are 4/48, or about 8%, quite small. Even if you have seen 27 other cards, and none of them are 7's, then your chances are still just 4/21, or 19%. You are "drawing to an Inside Straight," and your chances are not good.

On the other hand, if you were instead dealt a 5, a 6, a 7, and an 8, then either a 4 or a 9 would complete your Straight. This time, your chances of success are twice as good. This is called "drawing to an Outside Straight," which, as all good poker players know, is twice as likely to succeed as drawing to an Inside Straight.

So, while the poker players in movies are busy snarling and threatening and chewing tobacco, real poker players are carefully examining every card they can see, including opponents' "unimportant" discards. They are using this information to compute and update their probabilities of success, to make better decisions. It's true that psychological factors—things like bluffing and "tells" and keeping a "poker face"—are important, too. But probability is central to serious poker games, and players ignore it at their peril.

Showdown

You're playing five-card stud, and have three Queens in your first four cards. Meanwhile, your opponent has two 5's and one 4 showing, plus one secret face-down card. He is betting heavily, so you figure his face-down card might be another 5. Even so, three Queens beats three 5's, so you are not too worried. The only problem is that if his fifth card is another 5 or another 4, his hand will improve and will beat yours.

Your opponent plunks down $1,000 and challenges you to match it. What to do? Probably you will beat him, but what if his last card turns out to be a 5 or a 4?

You begin using probabilities. Even granting that your opponent's face-down card is a 5 (which it might not be), that still leaves 44 unseen cards (52 – 8). Of these, just four of them—the final 5, plus the three unseen 4's—will improve your opponent's hand. Right away that gives your opponent odds of just four out of 44, or 9.1%. Nothing to be too afraid of.

Furthermore, you notice that your "unimportant" fourth card (aside from your three Queens) is actually a 4! That reduces the number of unseen cards that can help your opponent from four to three. Now there are just three chances out of 44, or 6.8% probability, that your opponent can beat you. (And, even if he does get a 5 or a 4, there is a small chance that your last card will be a Queen or 4 to improve your hand as well, in which case you would still win.) It's practically in the bag.

Your opponent is glowering at you, trying to intimidate you into folding. But you cheerfully smile back and match his $1,000 bet (and perhaps even raise it). His final card turns out to be an 8, and you laugh all the way to the bank.

Once you understand the poker probabilities, what do you do with them? How do you decide whether to fold or call or raise the bet? This question is quite subtle, and entire books could be (and have been) written about the best decisions to make. However, a basic principle is that of *pot odds.*

Here's how it works. Suppose you have determined, as in the first Flush example above, that you have a 19% probability of winning. (This assumes that if you get your Flush you will win for sure, which isn't necessarily true, but let's assume it for now.) Suppose further that the bet is $10, and there is $300 already in the pot. (Of course, some of that $300 was originally your money, but that means nothing; at this point it belongs to the pot, not to you.) The question is, should you match the $10 bet or fold?

If you fold, you will neither spend that extra $10 nor get your hands on any of the $300 in the pot. End of story. If you stay in, then you must pay $10 right away. However, 19% of the time, you will win the entire $300 in the pot, plus the $10 you're adding. Since 19% of $310 is $58.90, this means that *on average,* by matching the bet, you will take in $58.90. Since $58.90 is a lot more than $10, it is in your interest to match the $10 bet and stay in the game. Of course, 81% of the time you will lose your $10. But the other 19% of the time, you will win such a large pot that overall it is worth the risk for you to stay in. In the long run, you will make more profit at poker by staying in when such situations arise than you will by folding.

By contrast, suppose instead that there is only $30 in the pot. In that case, *on average* you will win 19% of just $40, or $7.60, which is less than the $10 it would cost you to stay in. So, in that case, you should fold instead. It's all a question of pot odds.

If you keep track of probabilities and pot odds, you will make wiser decisions and will do well at poker over the long run. On the other hand, in certain poker games—from movie showdowns to *World Poker Tour*—players bet huge amounts on a single hand. Often they will go "all in," betting their entire holdings—perhaps a million dollars or more—on a single hand. Some people are impressed by such large bets, feeling that they show courage and strength and confidence, but not me. I figure anyone who bets all his money at once is trying to avoid the long run, and thereby avoid having the Law of Large Numbers reckon his true level of skill.

Know When to Hold'em

One currently popular version of poker is Texas Hold'em. Each player is dealt two face-down cards. Over the course of the hand, five additional cards are placed face-up in the middle of the table, for all the players to share together. The sequence is that there are first three face-up cards (the flop), then one more (the turn), and then a final one (the river), with a round of betting each time. Once the betting is all done, players who have not folded may choose any five cards from the seven they see (their two face-down cards, plus the five shared face-up cards). Occasionally, the five face-up cards will be best for everyone, resulting in a giant tie. But most of the time, the winning player will combine his own two cards with three well-chosen face-up cards.

Texas Hold'em is intriguing because, aside from the shared cards, all the other cards are face-down and secret. That makes it very difficult to guess what your opponents are up to, and whether or not they have a good hand. This encourages lots of bluffing and guessing and psychological games and so on—in fact, many serious players wear sunglasses in an effort to avoid being "read" or giving anything away. But even in Texas Hold'em, probabilities are crucial to long-term success.

As soon as the first two cards are dealt, good players begin assessing their odds. Higher cards (especially Aces and Kings) are better than lower ones; cards of the same suit (which might later combine to create a Flush) are better than cards of different suits; pairs are better than mismatched cards. It is too early to be confident, but already a sense of the probabilities can be gleaned.

Those who watch *World Poker Tour* will notice that even at this early stage, when each player has just two cards, the winning probability of each player is sometimes displayed on the screen. In other words, the television stations are able, upon seeing everyone's first two cards, to compute the precise probability of each player eventually winning the hand (assuming that no one folds).

How are these probabilities computed? It's not hard. All that is required is to run a computer program that considers every single possible deal of the remaining five face-up cards, and counts what fraction result in a win for each player. This may seem an impossible task, but actually it isn't. The total number of possible choices of five cards from a 52-card deck is 2.6 million; that sounds like a lot but a fast computer can cycle through the possibilities very quickly. Furthermore, once two players have each been dealt two cards already, there are only 48 cards left in the deck, and the total number of five-card hands is down to 1.7 million. So, in a matter of minutes or even seconds, tables can be generated showing the probability of victory for different combinations of two-card hands. And the probabilities in those tables are what gets displayed on the television screen.

Even without computers or tables or televisions, you can get a sense of your probability of victory. For example, suppose you are dealt two small Hearts. Since your cards are small, you might well be defeated by higher cards, or higher pairs, later on. However, if the five face-up cards contain three or more additional Hearts, you will get a Heart Flush and will probably win. So, what is the probability that you will get a Heart Flush?

You can see only your own two cards, and this leaves 50 unseen cards in the deck, of which 11 are Hearts and 39 are not. The total number of different combinations of five cards from 50 is 2,118,760, and each combination is equally likely. Different combinations include different numbers of Hearts, as shown in Table 4.1.

Table 4.1 Probabilities for Hearts on Table When You Have Two Hearts

# Hearts	# Combinations	Probability
0	575,757	27.2%
1	904,761	42.7%
2	502,645	23.7%
3	122,265	5.77%
4	12,870	0.61%
5	462	0.02%
Total	2,118,760	100%

To connect for a Heart Flush, you need three more Hearts. So, your chance of success is equal to the sum of the probabilities of the table being dealt three, four, or five more Hearts, which works out to 6.4%. Not very likely, and not worth fighting for if your opponents are betting high.

On the other hand, suppose you manage to stay in there long enough to see the first three shared face-up cards, and to your joy the flop has two Hearts out of three. Now things are more promising; you just need one more Heart in the last two face-up cards. What are your chances? You have now seen a total of five cards (two in your hand, plus three in the flop), of which four are Hearts. This leaves 47 unseen cards, of which nine are Hearts. The total number of possible two-card selections from a deck of 47 cards is 1,081. Of these, 36 consist of two Hearts, and a further 342 consist of one Heart. So, the total number of combinations that will give you a Heart Flush is 378, for a probability of 378/1,081, or 35%. Now you're talking! So, if you have four Hearts, and there are still two cards to come, your probability of getting a Heart Flush is 35%. If the pot odds are at all reasonable, you should stay in.

In this example, suppose the turn (the fourth face-up card) is not a Heart. Now you have just one chance left to connect for your Heart Flush. Of the 46 unseen cards, nine are Hearts. So at this point your chance of success is just 9/46, or 19.6%, a fair bit lower. You may wish to fold or you may not, depending on how the betting has gone, how many opponents are left, and how good you think your opponents' cards are.

Of course, it's difficult to compute all of these probabilities while sitting at a poker table. However, you can still use various rules of thumb to approximate the probabilities. Suppose again that you have four Hearts, and are hoping for at least one more Heart in the final two face-up cards (the turn and the river). You know that there are 47 unseen cards, of which nine are Hearts. So, the probability of a Heart on the next card is 9/47. Since there are two more cards to come, the probability of getting at least one Heart is roughly twice as high, or about 18/47. (In fact, 18/47 equals 38.3%, which is indeed close to the true probability of 35.0%. The error comes from double-counting the case where *both* cards are Hearts, which has a probability of 36/1,081, or 3.3%.) Now, 18/47 is less than one half but more than one third. So, your probability of getting another Heart is reasonably high, but is still less than a 50–50 proposition. By combining this observation about probabilities with a look at the pot odds, and an appropriate level of psychological gamesmanship, you can decide your next move (fold, call, or raise) with added confidence.

Simple probability calculations and a few rules of thumb can give you an idea of your hand's chance of success. Together with a consideration of pot odds, this thinking can significantly improve the poker decisions that you make. Probability theory cannot completely replace the bluffing and psychology of poker, but it sure can help.

A Big Blackjack Attack

Another popular casino game is blackjack, also called 21. The player repeatedly decides whether to take another card ("hit") or not ("stand"). Once the player stands, then the dealer also takes cards one at a time. At the end, the sum of the cards of the player and the dealer are compared (counting face cards as 10, and Aces as either one or 11 as appropriate). The winner is whoever's total sum is closest to 21, without going over. For example, if you have a King, a 6, a 3, and an Ace, your total is 20. If the dealer has a Queen, a 5, and a 4, then his total is 19. You win the round.

Blackjack rules vary slightly from casino to casino, but they usually include the following:

- You get to see the dealer's first card, even before making your decisions about hitting or standing.
- If neither you nor the dealer goes above 21, whoever has a higher sum wins the bet.
- A tie is a tie; if you and the dealer have the same point total, your bet is refunded.
- If you get blackjack (a sum of 21 on your first two cards, e.g., an Ace and a Queen), you are paid 1.5 times your bet, regardless of the dealer's cards (unless the dealer also gets blackjack).

- You may avail yourself of various other betting options, if you wish, such as: splitting pairs (if your first two cards are the same, you can separate them and play two different hands), doubling down (after your first two cards, you can double your bet and receive just one more card), insurance (if the dealer starts with an Ace, you can then make a side bet that their next card will be a 10 or face card), and surrender (where you give up right away and lose only half your bet).
- The dealer, meanwhile, has no choice at all, and must keep taking cards until the sum is 17 or higher, at which point he must stand. (In some casinos, the dealer still takes another card if he has a "soft 17" like Ace–6, or Ace–4–2.)

At first glance, these rules seem eminently fair. Whoever has the higher total sum (without going over 21) wins, and ties are treated fairly. Furthermore the player gets certain special advantages, like a special payout for blackjack, and extra betting possibilities (which are optional), while the dealer has no choice whatsoever and must play according to a prespecified strategy. All in all, it sounds like an advantage for the player, not the dealer. Yet casinos everywhere make a pretty penny from dealing out blackjack. How is this possible?

The catch, of course, is that if the player goes bust by exceeding 21, he loses regardless of the dealer's cards. To put it differently, if both sides go bust (or would have gone bust had the game continued), the dealer wins. That is the only rule in the casino's favor, but it is sufficient to keep the money rolling in.

While blackjack offers the player a number of choices, the main one is when to hit and when to stand. Obviously, if your sum is 11 or less, you want to hit. Or if it is 20 or 21, you want to stand. But what if your sum is, say, 15? In that case, if you hit, you might get a 5 or 6 and end up with a great hand. On the other hand, you might get a 7 through King and go bust. What should you do?

The basic principle is the same as in all other card games: all unseen cards are equally likely to appear next. However, there is one difference; casinos normally deal blackjack by shuffling many decks, six or more, all together and reshuffling often. Therefore, having seen a few previous cards will make almost no difference to the probabilities for what happens next. Furthermore, most casinos explicitly forbid the player to count cards that have passed by so far, and will expel players if they are caught making notes. (It is possible to use memory tricks to get a sense of the ratio of high to low cards remaining, and some experienced players do have some success this way, but it is less and less effective the more decks that are in use, and the more often they are reshuffled.) So for now we assume that in blackjack any card from Ace through King is equally likely to come up next, regardless of what cards have already been played.

Suppose you are playing blackjack in a casino (with many decks of cards), and you are dealt a Jack and an 8 (for a sum of 18). For your next card, of the 13 possibilities from Ace through King, 10 of them (all the 4's and above) will make you go bust. So, your probability of busting is 10/13, or 77%, which is far too high. It is best to stand at that point.

If you stand with a sum of 18, how will you do? The dealer must hit until he reaches 17 or more, regardless of your cards. Overall, the long-run probabilities for the dealer's final result (when playing with many decks) are shown in Table 4.2:

Table 4.2 Long-Run Probabilities for Blackjack Dealer

Dealer's Final Result	Probability
17	15.47%
18	14.76%
19	14.00%
20	18.50%
21	9.55%
Bust	27.73%

So, if you have a sum of 18, you will win if the dealer ends up with 17 or goes bust, with a total probability of about 43%. You will tie if the dealer also gets 18, with probability about 15%. And, you will lose if the dealer gets 19, 20, or 21, with total probability of about 42%. By stopping with a sum of 18, you have made the game virtually even; in fact, it is now ever so slightly weighted in your favor. This is certainly better than if you had hit with a 77% chance of immediately going bust.

In addition to these considerations, there is one other card that you should note carefully. That is the dealer's first (face-up) card, which has a big effect on the dealer's likely final result. In terms of this first card, the long-run probabilities for the dealer's final total (when playing with many decks) are shown in Table 4.3.

Table 4.3 Probabilities of Blackjack Dealer Results, Based on First Card

First Card	17	18	19	20	21	Bust
Ace	13.41%	13.41%	13.41%	13.41%	36.48%	9.89%
2	14.64%	14.03%	13.37%	12.66%	11.90%	33.41%
3	14.16%	13.59%	12.98%	12.32%	11.61%	35.33%
4	13.68%	13.15%	12.60%	11.97%	11.31%	37.32%
5	13.19%	12.70%	12.17%	11.61%	10.99%	39.33%
6	12.48%	12.03%	11.54%	11.01%	10.44%	42.50%
7	38.50%	9.51%	9.05%	8.56%	8.03%	26.34%
8	14.31%	37.39%	8.39%	7.94%	7.45%	24.52%
9	13.28%	13.28%	36.36%	7.36%	6.91%	22.82%
10/Face	12.31%	12.31%	12.31%	35.39%	6.39%	21.28%

Table 4.3 shows that the dealer's probabilities depend greatly on the first card. An Ace provides so much flexibility (since it can count as either one or 11) that the probability of going bust is reduced to less than 10%, and there is a good shot at 21 (since a 10 or face card might appear next). On the other hand, starting with a 6 gives the dealer a whopping 42.5% probability of going bust (if a 6 is followed by a 10 or face card, the dealer has 16, which is likely to go bust). By contrast, starting with a 7

gives a good shot at 17, starting with a 9 gives a good shot at 19, and so on, because getting a 10 or face card next is so likely.

How can we put this knowledge to use? Suppose you are dealt a Jack and a 3 (for a sum of 13), and the dealer's first card is a 5. Should you hit or stand? Well, if you hit, then you will go bust (and lose) if you get a 9, 10, Jack, Queen, or King, for a probability of 5/13, or 38.5%. Otherwise, you will improve your hand somewhat, but not necessarily enough to beat the dealer. Overall, if you hit just one more time, then your probability of winning depends on your next card, as shown in Table 4.4.

Table 4.4 Probabilities in "Jack–3 vs. Dealer's 5" If Hit Once

Next Card	Probability	Total	Prob. of Win	Prob. of Tie
Ace	1/13	14	39.33%	0%
2	1/13	15	39.33%	0%
3	1/13	16	39.33%	0%
4	1/13	17	39.33%	13.19%
5	1/13	18	52.52%	12.70%
6	1/13	19	65.22%	12.17%
7	1/13	20	77.39%	11.61%
8	1/13	21	89.01%	10.99%
9	1/13	22	0%	0%
10/Face	4/13	23	0%	0%
Overall:			33.96%	4.67%

Table 4.4 tells us that when the probabilities of winning based on various cards are all combined together, the overall probability of winning if you hit once is 33.96%, with a further 4.67% chance of a tie.

Meanwhile, the dealer started with a 5, which is not a good first card. We know from Table 4.3 that he now has a 39.33% probability of going bust, so if you stand you have a good chance (39.33%) of winning even with your paltry sum of 13.

This 39.33% chance of winning by standing is more than the 33.96% chance of winning (or even the 38.63% chance of winning or tying) by hitting once more. Thus, if you have a Jack and a 3, and the dealer's first card is a 5, overall you are better off standing than trying to improve your hand with a hit.

Serious blackjack players have done calculations such as these in great detail, running computer simulations and calculating many different probabilities. They have come up with a "basic strategy" to maximize the probability of success in various situations. This strategy has detailed rules for when to split, when to double down, when to hit, and when to stand. (Playing with this basic strategy reduces the house advantage to just over 0.5%, but even then you will still lose money in the long run.) The basic strategy includes the rule that if your sum is between 13 and 16 without any Aces counting as 11, and the dealer's first card is between 2 and 6, you should always stand. It's just like our Jack–3 example: if you hit you might improve your hand, but you might also go bust, and overall it is best to stand and hope that the dealer goes bust instead.

Patience, Patience

An understanding of randomness can increase your probability of victory at games of chance. However, the final requirement to becoming truly a winner is patience.

The Law of Large Numbers says that the person with the highest probability of winning will win the most in the long run. This doesn't mean you will win every time, just that you will win more than others if you play the game over and over again. Once you have figured out how to increase your probability of winning, you may need to play *a lot* before your winning ways click in. (Similar considerations apply to investing in the stock market: to be successful, it is not necessary that all of your stocks go up all of the time, just that they go up on average.)

Unfortunately, sometimes repetition is not an option, in which case all you can do is make your probability of winning as large as possible and take your chances.

The Portuguese Postal Puzzle (A True Story)

Last year, a researcher I had supervised moved back to his native Portugal. Shortly before he left, he hosted a delightful farewell party, complete with delicious Portuguese food.

As part of the festivities, he and his partner organized a simple contest: guess the postal code of their new address in Portugal. We were told that the postal code was somewhere between 1000 and 9999. Whoever guessed the closest would win a small prize.

A sign-up sheet was circulated, and guests wrote down their favorite number, or a number in the middle of the range, or some other arbitrary choice. But I vowed that, using the Probability Perspective, I could do something cleverer.

I decided to stand back and let the other guests choose their guesses first. Finally it was my turn. Looking over the list of guesses, I saw that there were none at all between 5000 and 7440. Aha, I thought, a nice large gap. By centering my guess on this gap, I would be far away from all the other guesses, thereby increasing my chance of winning. So I chose the number 6220, right in the middle of the gap.

I sat back, feeling smug. There were 20 guests at the party, so if I'd chosen my number arbitrarily, I would have just one chance in 20, or 5%, of winning the prize. But with my clever choice in the middle of a gap, I would win if the answer was anywhere between 5611 and 6829. Assuming all numbers between 1000 and 9999 to be equally likely, I now had a 13% probability of winning the contest—over twice as good as the 5% I started with. I had used the Probability Perspective to improve my odds.

After much delay, the results were announced. And . . . I lost! The prize was claimed instead by someone who had guessed on a purely arbitrary basis. In my crushing defeat, I consoled myself with the knowledge that if similar contests were held 100 times in a row, and I employed the same strategy each time, the Law of Large Numbers would kick in and I would win about 13 of the 100 contests—more than any other competitor.

Ever since that party, I have been desperately searching for 99 more Portuguese researchers, each holding a similar postal-code guessing game. If you know of any, please let me know.

To emerge victorious in games of chance requires three ingredients. First, you have to study the game carefully, to find a strategy that will win *on average*. Second, you have to repeat your strategy over and over again. Third, you have to wait patiently for the Law of Large Numbers to eventually carry you to victory.

5

Murder Most Foul

Measuring Trends

No event attracts as much public attention as killings. The more gruesome the act, and the more innocent-seeming the victim, the more attention we pay to what Shakespeare described as "murder most foul." More so than traffic accidents, disease, starvation, or even airplane crashes, murders strike a chord with people and make them fear that "it could have happened to me."

Our fascination with murder is not lost on the media, which constantly put homicide-related stories on their front pages—not just on the day of the event, but for weeks afterward, whenever there is an arrest or a new eyewitness or court appearance or other (perhaps minor) development. Nor is it lost on the entertainment industry, which features murder in a large percentage of movies and television programs.

Our interest is also not lost on the police themselves, who aren't shy about emphasizing violent crime in our communities, along with their need for additional resources to fight against it. "Gun-crazed gangsters terrorize at will," declared Toronto's police chief recently. "We're having difficulties coping with this gun-crazed, drugs-and-gangs mentality that prevails in our community."

Politicians sometimes stoke fear of crime for electoral success. The winning election platform in Ontario's 1995 election emphasized "the increasing incidence of violent crime," and opposition politicians chimed right in by declaring that the incidence of violent crime was increasing, and the level of violence was becoming almost indescribable.

So it seems that the media, the police, the politicians, and the citizenry all agree that violent crime is on the rise. But is it really so?

The Facts, the Counts, and the Rates

From a Probability Perspective, saying it is so doesn't make it so. Even if the media and politicians are playing up crime, that does not necessarily mean crime is on the rise. The only way to know for sure is to check the facts.

Data about homicide and other criminal activity are available from a variety of sources, including government agencies, police records, and public health bodies. In fact, the data from these different sources do not precisely agree; for example, some deaths are considered accidental at the time (and are recorded as such in health records) but later are ruled to be homicides. However, even a rough knowledge of actual data about criminal activities is still far more indicative of the true state of affairs than all the world's screaming headlines and blustering politicians and frightening feature films.

As with any data, it is important to use crime statistics correctly, and to understand the difference between *counts* and *rates*. For example, in the year 2000, there were 1,051 homicides in France and just 370 homicides in Lithuania. Aha, you might conclude, when it comes to homicides, Lithuania is a very safe country, while France is a more dangerous country. Right?

Actually, wrong. In the year 2000, the population of France was 59,225,683, while the population of Lithuania was just 3,620,756, over 16 times smaller. So the rate of homicides in France in 2000 was 1,051 homicides per 59,225,683 people, or one homicide for every 56,352 people. Meanwhile, the rate of homicides in Lithuania in 2000 was 370 homicides per 3,620,756 people, or one homicide for every 9,786 people, a much higher figure.

In other words: In 2000, a (randomly chosen) person living in Lithuania was nearly six times as likely to be a victim of homicide, as was a (randomly chosen) person living in France. What this example shows is that it is not meaningful to consider total numbers of homicides. Rather, we must consider the rate; that is, we must divide the number of murders by the population involved.

Such rates are usually written as the number per 100,000 people. Thus, France's homicide rate in 2000 is equal to 1,051 homicides divided by 59,225,683 people multiplied by 100,000 (1,051 ÷ 59,225,683 × 100,000), which works out to 1.78. By contrast, the corresponding rate for Lithuania is 10.22. Figures for a few other countries are shown in Table 5.1; note that the countries with the largest number of homicides do not necessarily have the highest rate of homicide.

Table 5.1 Homicide Counts and Rates for Selected Countries, Year 2000

Country	Population	# Homicides	Rate per 100,000
Lithuania	3,620,756	370	10.22
Estonia	1,431,471	143	9.99
United States	281,421,906	15,980	5.53
Scotland	5,062,900	104	2.05
Australia	19,360,618	363	1.87
Canada	30,689,035	546	1.78
France	59,225,683	1,051	1.78
England & Wales	52,140,200	679	1.30
Germany	82,797,000	961	1.16
Japan	126,550,000	1,391	1.10

The same is true when we compare cities or states or any other jurisdictions. For example, in the three-year period 1998–2000, there were 538 homicides in London, England, and just 89 in Amsterdam, Netherlands. But the population of London is over 7 million, while fewer than 750,000 live in Amsterdam. When these homicide counts are converted into annual rates per 100,000 people, this corresponds to 4.09 for Amsterdam and just 2.38 for the safer city, London. Similarly, during the period 1998–2002, New York State averaged 930 homicides per year, while South Carolina averaged just 282. However, nearly 19 million people live in New York State, while just under 4 million live in South Carolina. So the corresponding homicide rates per 100,000 are 4.97 for New York State and 7.07 for South Carolina. For Massachusetts, it's 2.19. In other words, if homicide is your greatest fear, you're significantly safer in New York than in South Carolina, and over three times safer in Massachusetts. Many people find this statement surprising. We have been conditioned to think that high-population regions are unsafe, and that more total homicides mean more danger. But it ain't necessarily so.

The lesson here is that when we compare quantities for different populations—whether homicides, or the amount of beer consumed, or the number of automobile accidents, millionaires, leisure suits, or Nobel Prizes—the only meaningful way to make an evaluation is to calculate a rate. Comparing the total counts among entities with different populations is completely misleading.

Homicide Trends

People's greatest fears often involve a perceived increase in the crime rate. We tend to accept things the way they are, but we hate the thought that they might be getting worse. How can we tell if the homicide rate is increasing or decreasing?

Let's begin by considering the number of homicides in the United States, by year, from 1960 to 2002 (using raw data from the U.S. Department of Justice). One strategy is simply to make a list of the homicide counts, by year, as follows: 9110, 8740, 8530, 8640, 9360, 9960, 11040, 12240, 13800, 14760, 16000, 17780, 18670, 19640, 20710, 20510, 18780, 19120, 19560, 21460, 23040, 22520, 21010, 19310, 18690, 18980, 20610, 20100, 20680, 21500, 23440, 24700, 23760, 24530, 23330, 21610, 19650, 18210, 16974, 15522, 15586, 15980, 16204.

Such a list purely of numbers is hard to interpret. However, we can already notice a few trends. The number of homicides starts relatively small (fewer than 10,000) in the 1960s, climbs up toward a peak of over 24,000, and gradually decreases toward the end.

It is much more informative to view the same figures in a graph, as in Figure 5.1. This graph appears to confirm our original suspicion, namely that there was a significant increase in the homicide counts until the mid-1970s, and then a gradual levelling off followed by a slight decline.

Figure 5.1 Homicide Count: United States (by year)

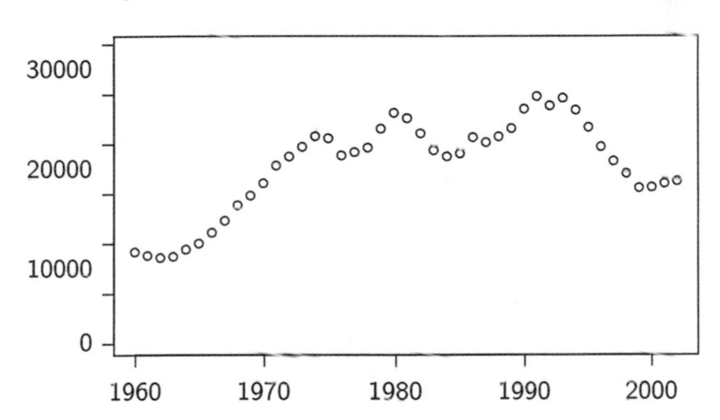

But wait. We have already seen that what really matters is the rate of homicides per 100,000 people (say), not the pure counts. Indeed, the population of the United States increased from about 180 million in 1961 to about 210 million in 1975 to over 280 million in 2002. Figure 5.2 shows a graph of American homicide rates 1960–2002. This graph is similar to the previous one, and confirms our earlier impression. The increase from 1960 to the mid-1970s is still there, as is the later decrease. However, the rate of decrease from the mid-1970s onward is now larger because the increase in population during that time period has been accounted for.

Figure 5.2 Homicide Rate per 100,000: United States (by year)

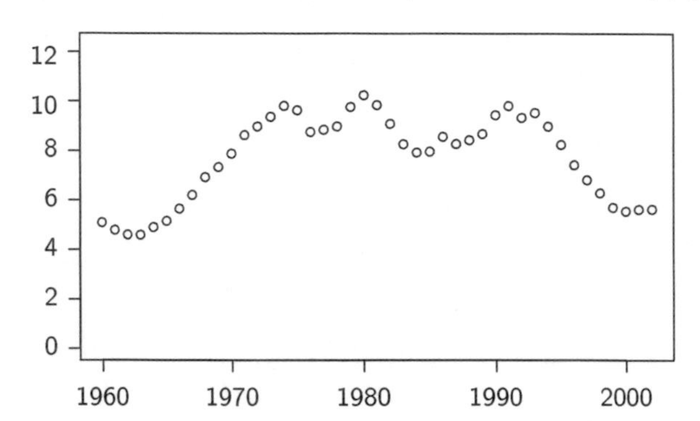

On the other hand, there are a few years around 1980, and a few more around 1990, which do not appear to fit into the general trend. What are we to make of them? Do they change the entire analysis, or do they just signify temporary, random fluctuations? How can we really tell whether or not the homicide rate shows a decreasing trend from the mid-1970s until today and, if it does, how large is the decline?

Measuring Trends: Regression

It is all too easy to draw false conclusions about trends based on just a few observations. To measure trends more accurately, statisticians use a technique called *regression*. In its simplest (linear) form, regression provides a simple rule for determining the "line of best fit," which is the line that best approximates the increasing or decreasing trend of the values being considered. It describes a mathematical formula for determining the line that comes closest to the observed data values. (More precisely, the line minimizes the sum of the squares of the distances between the line and the data.) The formula is specific and unambiguous, thus avoiding any bias or subjectivity that might arise from drawing "as good a line as you can" just by looking.

Weight Gain, Weight Loss

This year you're really going to lose some weight! You've got a new diet and a new exercise program. Oh sure, you don't always stick to your goals, but still you are feeling hopeful.

The first day you weigh yourself: 170 pounds. Too heavy. But you remain confident.

The second day you weigh yourself again: 172 pounds. Not a good start. Hang in there.

The third day you're up to 174 pounds. Rats!

But the fourth day, that glorious fourth day, you weigh in at 173 pounds. Aha, your diet is really working now. You're losing one pound a day. At this rate, in one month you'll be down to 143 pounds, and as slim and healthy as can be.

Let us again examine the U.S. homicide rate per 100,000 people, this time just for the years 1975–2002 (the period of apparent decrease), together with a linear regression line of best fit:

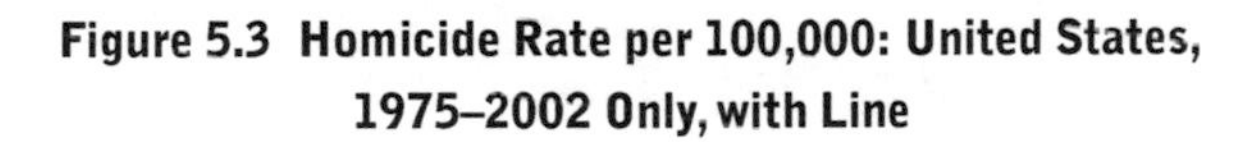

Figure 5.3 Homicide Rate per 100,000: United States, 1975–2002 Only, with Line

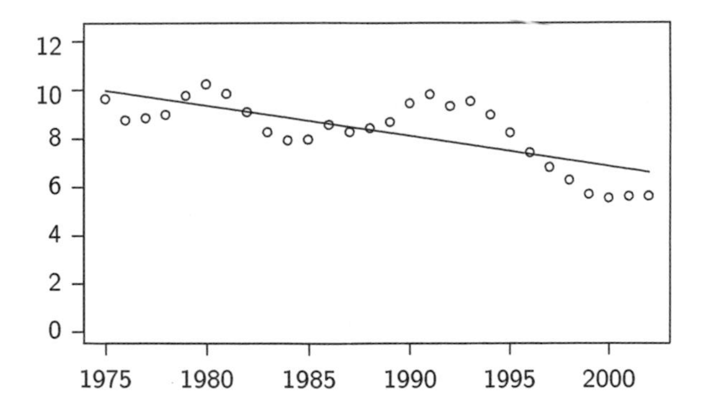

This line has a downward slope of 0.126. This means that, on average over this time period, the annual U.S. homicide rate was *decreasing* by about 0.126 homicides per 100,000 people per year. Of course, that is not a very *large* decrease, but it is definitely a decrease. In short, the homicide rate in the United States was not going up between 1975 and 2002. In fact, since the mid 1970s, this rate has most definitely being going down slightly.

This trend is most pronounced in the years since 1990. For that time period, the U.S. homicide rate was decreasing quite significantly, with a downward slope of 0.454 per year, as is evident in Figure 5.4. In other words, since 1990, the U.S. homicide rate has definitely been going down, at the same time as many U.S. media outlets and politicians were hyping fears of murderers everywhere.

Figure 5.4 Homicide Rate per 100,000: United States, 1991–2002 Only, with Line

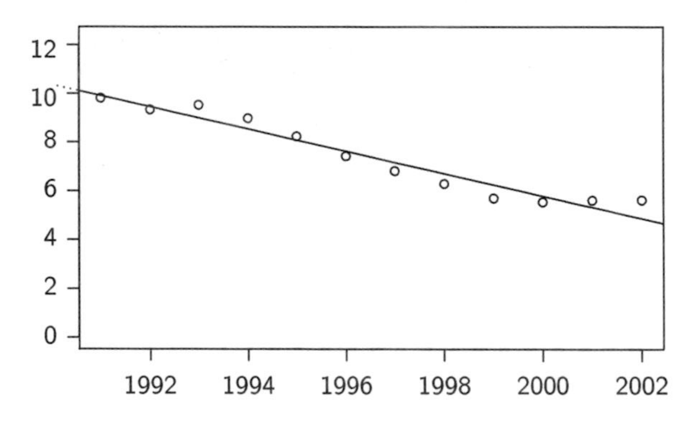

The fact that the U.S. homicide rate was decreasing throughout the 1990s runs counter to many people's impressions. So perhaps other violent crime increased during the same period? Hardly. In fact, U.S. Department of Justice data show that the annual rate of violent crime per

100,000 people in the United States since 1990 has been decreasing quite significantly, by about 28.7 per year:

Figure 5.5 Violent Crime Rate per 100,000: United States, 1991–2002, with Line

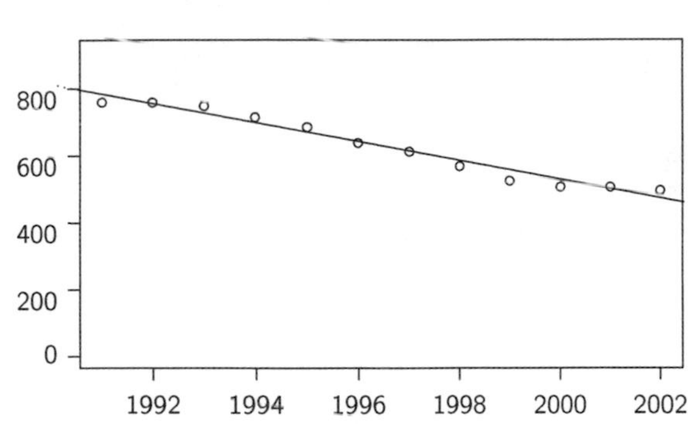

So what have we learned?

- To determine trends, it is best to use a regression, such as a line of best fit, rather than to focus on just a few individual yearly totals that could be deceptive.
- The trend that you determine depends on how you analyze the numbers (for example, counts versus rates, what geographic region you include, which years you include). When you see any statistical analysis, always read the fine print.
- In the case of the homicide rate for the United States, no matter how you do the statistical analysis, the only possible conclusion is that the homicide rate in recent years has been decreasing, not increasing. Absolutely 100% for sure.
- Just because the media, the politicians, and the police all say the same thing, it isn't necessarily true. As the rap group Public Enemy would put it, "Don't believe the hype."

Murders the World Over

What about other countries? The Canadian homicide rate is about one third that of the United States, and it has also been decreasing slightly (at about 0.042 per year) since the mid-1970s (using raw data from Statistics Canada):

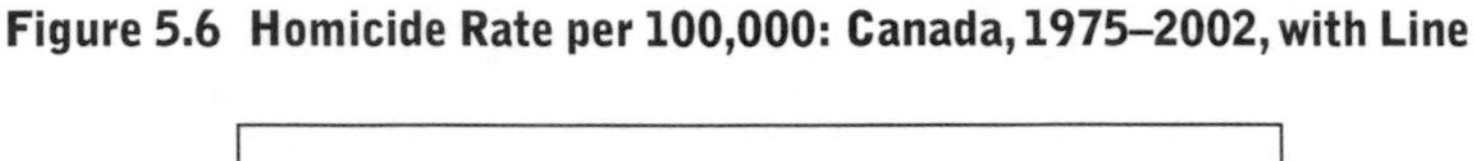

Figure 5.6 Homicide Rate per 100,000: Canada, 1975–2002, with Line

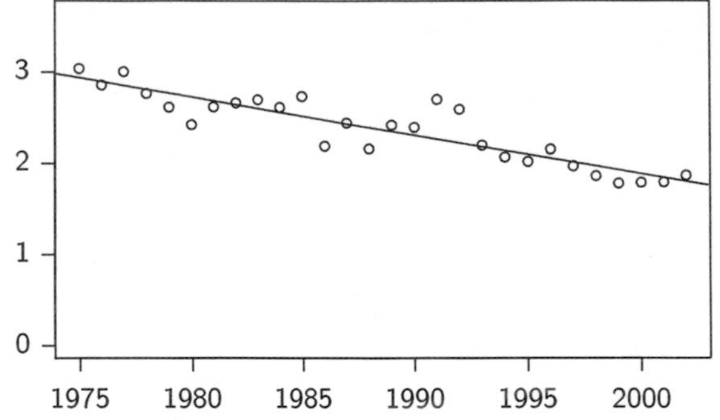

Once again the effect is most pronounced in more recent years, where it is decreasing at a rate of about 0.074 per year:

Figure 5.7 Homicide Rate per 100,000: Canada, 1991–2002, with Line

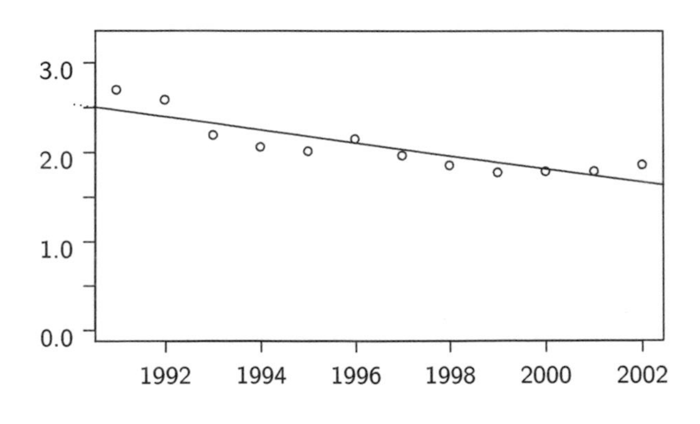

For Australia since 1990, the homicide rate has increased some years and decreased others, but overall stayed almost exactly constant, with a tiny downward slope of just 0.0006 (using raw data from the Australian Institute of Criminology):

Figure 5.8 Homicide Rate per 100,000: Australia, 1991–2002, with Line

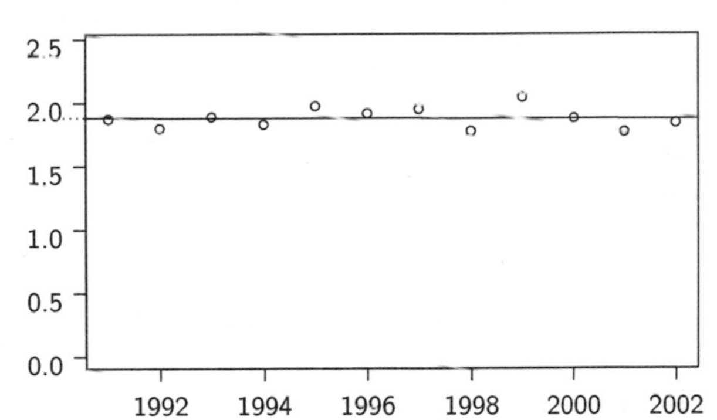

As for Great Britain (England, Scotland, and Wales), the homicide rate is lower than that in the United States, Canada, or Australia, but since 1990 it has actually gone up slightly, increasing by about 0.025 per year (using raw data from the British Home Office and the Scottish Executive):

Figure 5.9 Homicide Rate per 100,000: Great Britain, 1991–2002, with Line

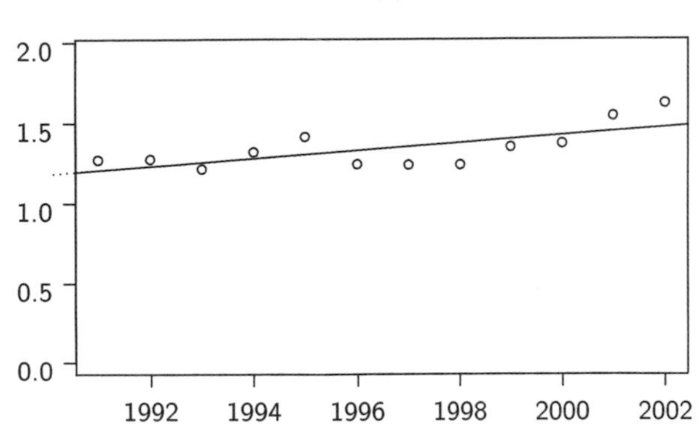

Table 5.2 is a summary of trends in these four countries between 1991 and 2002.

Table 5.2 Homicide Rates (per 100,000) and Trends, 1991–2002

Country	Average Rate	Trend per Year
United States	7.40	Decreasing by 0.454
Canada	2.05	Decreasing by 0.074
Australia	1.87	Decreasing by 0.0006
Great Britain	1.34	Increasing by 0.025

It is perhaps heartening to note that the highest homicide rate (for the United States) is also decreasing the fastest, while the only homicide rate that is increasing at all (Great Britain's) is very low to begin with. In any case, the numbers make clear that in contrast to popular impressions and media hype, murder rates are quite low and are generally decreasing.

Furthermore, rates are not the only issue in which reality differs from media portrayals; another concerns victim–offender relationships. When people fear murder, they generally fear being killed randomly by a stranger. But most murders do not occur that way. According to a Statistics Canada study of 15,163 homicides between 1974 and 2002, in only 15.5% of the solved homicides was the perpetrator unknown to the victim, a figure that has also been decreasing slightly since 1975. By contrast, about 18% of solved homicides were committed by the victim's spouse. Similar numbers were obtained by an Australian Institute of Criminology study of 2,757 homicides in Australia between 1989 and 1996, and (with some differences) by an FBI study of 14,408 homicides in the United States in 2003. The figures are shown in Table 5.3.

Table 5.3 Victim–Offender Relationship Percentages (Solved Homicides)

Relationship	Australia	Canada	United States
Total Family	35%	37.7%	22.6%
Spouse	20.3%*	18.4%	8.7%
Parent	6.7%*	8.31%	2.93%
Other Family	8.0%*	10.94%	10.9%
Other Intimate	*	3.53%	7.8%
Acquaintance	38%	43.2%	47.2%
Stranger	16%	15.6%	22.4%

* The Australian study puts Other Intimate in with Spouse, and puts Child in with Parent.

These figures provide a Probability Perspective on high-profile, terrible events like the abduction and decapitation of a defenseless child by a psychopath. Such horrors remain in the front-page headlines for months not only because they are so shocking, but also because they are very, very rare.

While the precise numbers vary somewhat from country to country, the conclusion is always the same. Contrary to the impressions created by the news media and by the movies, murder is very rare, and far more murders are committed by acquaintances and family members than by strangers. The next time you start worrying about crazy killers lurking behind every lamppost, perhaps you should check up on your spouse's recent movements instead!

Playing with Numbers?

These modest homicide rates do not square with the extreme statements of certain politicians and media representatives. So how do these people respond when they are directly confronted by the facts? Often with pure denial. A municipal politician in Toronto simply declared, "I don't agree that crime is down right across the board." And the police chief insisted, "The business of crime being down is really a facade. It's how you play with numbers. The people that are talking about crime [being] down must be reporting on some other community, not this one."

When pressed, these same people will point to people's *fear* of crime, as if to prove the validity of their point. I recently saw a television host ask a panel what should be done about the increasing crime rate. When several panelists responded that the crime rate is actually decreasing, not increasing, the host retorted, "Oh, come on now. Are you denying that there is an increasing *fear* of crime?" Or, as one politician frankly put it, "Don't just look at the statistics, look at the fear factor."

A politician once insisted that "people can *feel* that crime is increasing." Hardly. What people can "feel" is the extent to which fear of crime is played up by the media and the politicians. The true facts must be researched carefully and analyzed statistically, not felt.

The sad fact is that because so many people benefit from the public's fear of crime, it will not disappear easily. Media need this fear to sell newspapers; the entertainment industry needs it to market movies; politicians need it to get elected; and police need it to increase their funding. So I was proud of Toronto Mayor David Miller when he declared during a recent election campaign, "The truth is that Toronto is a safe city. Leaders don't go out and scare people to get elected." I was equally pleased when local journalist Joseph Hall systematically debunked the hype about increasing crime, concluding that "despite recent dire warnings . . . crime is falling." Or when columnist Doug Saunders noted that "by almost any gauge, the danger of terrorism, of crime, even of accidents is at a lower ebb than at almost any point in the last generation. What is dangerous about these times, in truth, is our fear." Hear, hear.

Does the decreasing crime rate mean that we should no longer worry about crime? That we no longer need police officers? Of course not. As long as there is any crime at all (which surely means forever), we need a strong and professional police force to protect us from harm. In addition, I have no doubt that our justice system could do with improvements to better serve crime's victims, keep guns off the street, and keep us safe from hardened criminals. On a personal level, if ever I or a loved one were a victim of violence, then of course I would want all the police protection and assistance I could get.

Furthermore, just because the crime rate has been decreasing in recent years, there is no guarantee that this decrease will continue. In the decade ahead, crime rates just might rise again.

On the flip side, even if it turned out that crime were increasing (which it isn't), we should still remember that violent crime is much less likely to harm us than, say, traffic accidents or disease. Thus, even while we remain vigilant, we should keep our fear in check—regardless of the crime trends.

In a sense, the decrease in the crime rate really isn't that important. Regardless of the numbers, there are no easy answers to the questions of how to balance control of crime with personal freedom, or police services with limited city budgets. It is not up to statistics or statisticians to decide what our priorities should be. What we *can* say is that such decisions should be made in the context of the true facts, not of exaggerated fears or misleading claims. If a politician or police spokesperson says that despite falling crime rates, we should still spend more money on policing, their argument is worthy of consideration and debate. But if a politician or police spokesperson claims that because crime is increasing we should spend more money on policing, he or she is simply not telling us the truth.

To quote *Dragnet*'s Sergeant Joe Friday, "Just the facts, ma'am."

6

Utility Functions

How to Make Decisions

Often we are forced to assess randomness when making decisions. Should we take an airplane even though it might crash? Should we buy a lottery ticket even though we might not win? Should we take out an insurance policy even though we might never collect? Should we go for a bike ride even though it might rain?

Of course, there is no magic solution, and some decisions will remain difficult, come what may. However, a bit of Probability Perspective and a few rules of thumb can help make many decisions easier. And for the really tough decisions, a utility function can help to sort out competing objectives.

Ignore the Extremely Improbable

The first rule when making decisions about randomness is that events of extremely small probability should generally be ignored. This is a very simple rule that most people do not follow.

A good example is lottery jackpots. Billions of dollars are spent on lottery tickets worldwide, by optimistic souls hoping to win a huge jackpot and live happily ever after. Is this a wise decision? Leaving aside the question of whether winning a lottery jackpot truly leads to happiness (it often has the opposite effect), what is the probability that you will win?

A typical commercial lottery might involve, say, selecting six different numbers between 1 and 49. If your six numbers match the six numbers later selected by the lottery company, you win (or share) the big jackpot. For such a lottery, the probability of winning the jackpot is one chance in the total number of ways of choosing six different numbers out of 49 choices, which is about one chance in 14 million. (The calculation is similar to the case of matching all the balls in a game of keno.) This is an extremely small probability. To put it in context, you are over 1,000 times more likely to die in a car crash within the year. In fact, you are more likely to die in a car crash *on your way to the store to buy your lottery ticket* than you are to win the lottery jackpot. Indeed, if you bought one ticket a week, on average you would win the jackpot less than once every 250,000 years. Furthermore, as the number of choices increases, the probability goes down even further; when picking seven numbers between 1 and 47, the probability of matching is one chance in 63 million. It may be true that *someone* is going to win the lottery jackpot this week, but let me assure you: that someone will not be you.

From a practical point of view, this means that when you are deciding whether or not to buy a lottery ticket, the possibility of winning the jackpot should not be a factor in your decision. Go ahead and buy lottery tickets if you find the experience entertaining or interesting or uplifting or fun. But you simply should not buy them with hopes of a jackpot victory. (I have never purchased a commercial lottery ticket; I know the odds too well.)

On occasion, lottery jackpots will grow to enormous size, perhaps hundreds of millions of dollars. It is tempting to buy a ticket then, because even though the probability of winning is so small, the payoff is so large. However, the larger the number of people who buy tickets, the greater the probability that you will have to share the jackpot even if you do win it. In such circumstances, it is better to choose unusual lottery numbers (best is a random choice; worst is 1–2–3–4–5–6 or your child's birthday) to reduce the risk of sharing. But the jackpot probabilities are so unimaginably small that, as a computer said about nuclear war in the Matthew Broderick movie *War Games*, the only way to win is not to play.

The principle of ignoring the extremely improbable applies much more widely. Franklin D. Roosevelt may have exaggerated slightly back in 1933 when he claimed that "the only thing we have to fear is fear itself." But it is certainly true that humans often worry unnecessarily about events of very small probability, which causes us to make poor decisions and to suffer stress and unhappiness.

For example, some friends of mine traveled to Israel in 2002, shortly after a series of high-profile terrorist attacks on civilians. Many people reacted with horror, implying that it was crazy to even *consider* traveling anywhere so dangerous. One acquaintance declared that going to Israel was just as foolish as driving the wrong way on a major highway.

I decided to check the facts. I discovered that during the period of heightened terrorist attacks from October 2000 through April 2002, there were a total of 319 people killed by terrorist attacks in Israel, about one person in 20,000. By comparison, about 750 Israelis died in motor vehicle accidents during this same time period. Thus, an Israeli was over twice as likely to die in a car crash (even without driving the wrong way on a highway) as to die in a terrorist attack, even during this period of increased terrorist activity. I explained these facts to my concerned acquaintance, but I didn't manage to convince him.

At about the same time, professional statisticians were making plans for the July 2004 annual scientific meeting of the Institute of Mathematical Statistics. This meeting was originally planned for Israel, but they decided to move it to Spain, out of fear of terrorism. Even among statisticians, the fear of exposing their colleagues to harm overwhelmed their Probability Perspective.

Overestimating the probability of highly improbable events can have serious consequences. For example, in the spring of 2003, a number of people living in the Toronto area contracted Severe Acute Respiratory Syndrome (SARS), a serious and potentially fatal viral infection. This outbreak was heavily publicized by the media, with numerous front-page headlines and television reports played around the world. But the total number of Toronto SARS fatalities through the entire duration of the crisis was fewer than 50. By comparison, about 1,000 Canadians die each year from common influenzas. A traveller visiting Toronto, even at the height of the SARS outbreak, was about as likely to die of influenza as of SARS, but I cannot recall any front-page headlines about an influenza outbreak, nor do I know of any tourists who changed their travel plans or behavior patterns to avoid contracting influenza. The SARS crisis caused the number of tourists visiting Toronto (and even the rest of Canada) to decrease dramatically, costing the city and the country billions of dollars, for no logical reason.

Table 6.1 Percentage of Deaths from Selected Causes, 2001

Cause	United States	Canada
Total deaths	2,416,425 (100%)	219,114 (100%)
Cardiovascular diseases	922,334 (38.2%)	74,824 (34.1%)
Cancer (all forms)	553,768 (22.9%)	63,774 (29.1%)
Lung cancer	156,058 (6.46%)	16,558 (7.56%)
Respiratory system diseases	230,009 (9.52%)	17,585 (8.03%)
Transportation accidents	47,288 (1.96%)	3,032 (1.38%)
Intentional self-harm	30,622 (1.27%)	3,688 (1.68%)
Fatal falls	15,019 (0.62%)	1,727 (0.79%)
Poisoning	14,078 (0.58%)	955 (0.44%)
Homicides (all)*	15,980 (0.65%)	553 (0.25%)
Homicide by relative	3,611 (0.15%)	208 (0.10%)
Homicide by stranger*	3,580 (0.15%)	69 (0.03%)
Homicide by spouse	1,390 (0.06%)	109 (0.05%)
Drowning and submersion	3,281 (0.14%)	278 (0.13%)
Smoke, fire, and flames	3,309 (0.14%)	243 (0.11%)
9/11 terrorist attacks	3,028 (0.13%)	—
Commercial aircraft*	275 (0.01%)	2 (0.0009%)
Lightning (year 2000)	50 (0.002%)	3 (0.0014%)

*Excluding 9/11 attacks.

(Figures from National Vital Statistics Reports and from Statistics Canada, and for the last two rows airdisaster.com and nationmaster.com; unknown victim–perpetrator relationship percentages extrapolated from known ones.)

Table 6.1 shows the percentage of deaths that resulted from various causes, in Canada and the United States in 2001; similar percentages apply in most years to most industrialized countries. Clearly, deaths from naturally occurring diseases (especially cardiovascular, cancer, and respiratory) are far more common than deaths from external causes. Even among externally caused deaths, transportation fatalities are far more common than deaths from homicide, airplane crashes, drownings, fires, or lightning. For example, the 3,000 people killed in the horrific 9/11 terrorist attacks represented just 0.13% of all American fatalities that year, corresponding to one of every 94,000 living Americans. This figure is equivalent to just over three weeks' worth of transportation accidents, meaning that the probability of a random American being killed by terrorism in 2001 (or, for that matter, during the entire period 2000–04) was about the same as that of being killed in a car accident in a single three-week period. That fact should not in any way minimize the outrageousness of these attacks, or the tragedy of the resulting deaths. However, it does provide a Probability Perspective on the numbers, and suggests that even the terrible 9/11 attacks did not significantly change the probability of sudden, unexpected death in the Western world.

If you are worried about dying, you should exercise and eat well to avoid cardiovascular disease, and quit smoking to avoid lung cancer. It makes far more sense to take care of your health than it does to worry about being murdered—to say nothing of deaths from terrorism, airplane crashes, drowning, burning, or lightning. And even if you do worry about murder, it makes more sense to fear a family member than a stranger. Those are the true facts, based on probabilities. (Of course, it is true that homicide and transportation accidents and terrorism tend to strike younger victims than do cancer and cardiovascular disease, which arguably makes them more tragic, but that fact is still dwarfed by the huge difference in the raw numbers.)

Despite these facts, the media focus far more on murders from unknown "bad guys" than they do on diseases and car accidents. Indeed, in the aftermath of the 9/11 attacks, North Americans' consumption of anxiety medications increased significantly, and everyone was talking about the fleeting nature of existence, how we have to live for the moment, and how vulnerable we all are. The equivalent number of deaths that occurs every three weeks in transportation accidents does not have the same effects.

Why do people fear terrorism and SARS so much more than automobile accidents and cardiovascular disease? Because terrorism and SARS seem new and unknown, and therefore uncertain. Humans can accept significant danger and loss of numerous lives, provided these happen in a manner to which they are accustomed. But when unexpected dangers emerge, they fear these more than is truly justified. *The Simpsons* got it right in a show where young Lisa Simpson was hanging out with a lowlife but harmless saxophone-playing street person, and her mother Marge drove up and whisked her away, saying to the vagrant, "It's nothing personal—I just fear the unfamiliar."

It may seem cruel or heartless to dismiss very small probabilities when the fear of harm relates to one's child or loved one. However, such feelings do not stand up to scrutiny. Most of us wouldn't hesitate to invite our sister to drive across town for dinner or a movie, yet about one person in 10,000 dies each year on our roads. The average person does the equivalent of driving across town at most twice a day, so your sister has at least one chance in 7 million of dying on her trip to see you. Is it heartless to issue such an invitation? Of course not. Rather, extremely small probabilities are, and must be, ignored every day.

The principle of ignoring the extremely improbable can resolve many other dilemmas as well. On a break from a recent bike ride in downtown Toronto, I found myself relaxing directly under the imposing 553-meter-high CN Tower, the world's tallest free-standing structure. Looking way up, I could make out the famous glass-floor section of the observation deck, 113 stories high. If that floor happened to break, people and glass shards would come hurtling straight toward me, too fast for me to escape, and would kill me instantly. I started backing away nervously, to take my break elsewhere.

Suddenly I stopped myself. *What was I thinking?* This glass floor has been in place for over 10 years, which is more than 5 million minutes. Even without taking into account the safety of its engineering and construction, the probability that it would collapse in the next minute must be less than one chance in 5 million. It simply wasn't worth moving out of the way or altering my plans in any way, based on such exceptionally small probabilities. (If the glass floor had been installed yesterday, I might have reconsidered.) I finished my break in peace, without moving anywhere, and without getting crushed to death either.

The Noise in the Night

After a long day at the office, battling traffic, running errands, and cooking dinner, finally you lie down in bed. You relax for a few minutes by reading a fascinating book about probability theory, and then you are ready for a nice, long sleep.

As you lean over to put down your book and switch off your lamp, you hear a sudden creaking noise. What was that? Who is there? Staying perfectly still, you listen carefully but don't hear anything further.

Now you're not sure. Was that noise a sign that a burglar has broken into your house? Should you phone the police? Should you grab your baseball bat? Should you go downstairs and investigate?

You think quickly. Your house hasn't been broken into for many years. What are the odds that a burglar happened to make a noise just at the exact moment when you were reaching over to put down your book?

On the other hand, your house is old, and a shifting weight on a bed is just the sort of thing that might cause the floor to creak a little. Perhaps putting down your book was the cause of the noise, rather than coinciding with it by chance.

Even in your tired state, you can see that it is far more likely that your movement caused the floor to creak than that a burglar arrived at precisely the same moment. Satisfied, you put down your book, switch off the lamp, and drift off happily.

Randomness and Indifference

Ignoring the extremely improbable is a sound, rational way to approach decisions, but if we take it to extremes, we might be tempted into recklessness or negligence.

Should we bother wearing a seatbelt in the car when the probability of getting into an accident on a 10-minute drive is so small? My answer is yes: I always wear a seatbelt (and a bicycle helmet), and not just because they're legally required. How do I justify this practice, given everything I have said so far in this chapter?

First, over the course of your life, you will probably take many different car and bike trips. The probability of being in some sort of accident—not necessarily fatal—over the course of so many trips is not inconceivable, and hence cannot safely be ignored. Forgetting a seatbelt once or twice probably isn't such a big deal, but never using it is asking for trouble.

Second, wearing a seatbelt or helmet requires little effort. It does not require cancelling a trip, or avoiding a fun event, or travelling great distances, or spending lots of money. So, even though an accident on any one particular excursion is very unlikely, wearing a seatbelt is such a simple action that it is worth the trouble. From this point of view, we should wear a seatbelt not because it is wise or moral or proper, but because it is easy.

Parents should perform a similar balancing act. On the one hand, it is fairly unlikely that your child will, say, seriously injure herself with a pair of scissors. On the other hand, the probability is not *so* low, especially if she uses scissors often, and simple precautions can help to prevent this risk. An admonition to your child not to run while holding scissors is definitely worthwhile.

Deciding that a probability is too low to bother responding to can be harmful on a societal level as well. Should we vote even though it's extremely unlikely that our one vote will make the difference? Should we bother to recycle or to avoid littering when one more piece of garbage has little impact? Should we bother to conserve energy when it is inconceivable that our own personal usage will push the planet past its tolerance?

Of course, if many people vote, we will (we hope) get more representative government. If many people recycle, we will create a cleaner and greener world. And if many people conserve energy, our lifestyle will be more sustainable. Philosophers refer to this issue as "individual versus collective rationality": if many people take an action, it benefits us all, but if just one person takes the same action, it inconveniences him or her without having much chance of benefiting anyone.

So why should we bother? Part of the answer is that we hope that our actions will inspire others to act similarly, and that many people's actions taken together will be significant. Unfortunately, we can't always be confident that our actions will inspire others: maybe no one will see us vote or recycle. Should we persevere with our actions anyway?

I think Jean-Paul Sartre got it right when he said that our choice of actions dictates the way we want everyone else to act, too. "In creating the man we want to be," he wrote, "there is not a single one of our acts that does not at the same time create an image of man as we think he should be." In other words, by our very act of voting or recycling, we are saying that we think everyone else should vote and recycle, too.

We should ignore extremely unlikely events when it comes to worrying about murder and terrorism, or to wasting money on lottery tickets, but we shouldn't let small probabilities keep us from following simple safety measures like seatbelts, or from taking positive actions like voting and recycling, in the hopes that others will follow our lead.

Maximize Your Average Happiness

It is all well and good to talk about ignoring events that are extremely improbable, but often we are confronted by choices involving randomness that are not unlikely at all.

Walk or Ride?

You overslept, and now you're rushing to a 9:00 A.M. meeting. It's 8:50, and your office is still a 15-minute walk away at maximum speed. You're going to be five minutes late.

You consider your options. You'll never get a taxi at this hour. However, there is a bus that could take you to your office in five minutes flat. If you wait at the bus stop, and if the bus arrives within five minutes, you will be on time. Great!

Unfortunately, the bus is erratic. Sometimes it arrives right away, but sometimes it doesn't show up for 20 minutes. If you wait for the bus and are unlucky, you might end up being as much as 15 minutes late for your meeting, instead of just five. Should you risk it? Should you wait for the bus, to have some chance of arriving on time, but at the risk of possibly being much later? Or, should you continue walking and be certain to arrive precisely five minutes late?

You recall that you are on probation at your job and that your stern and punctual boss will surely fire you if you are late. You decide to wait for the bus, since your only hope is that the bus arrives quickly and gets you to work on time. Fortunately, the bus is waiting, and your job is saved.

That evening you are meeting a friend for drinks at 9:00 P.M. Once again, you find yourself running late, and once again you have to choose between slight but certain tardiness from walking, or a bus with possible punctuality but possible major delays. Being a few minutes late for drinks is no big deal, but if you are very late your friend might get concerned or leave. You decide that walking is your best option.

Making a choice like walking versus taking the bus depends upon your feelings about the various consequences of your being on time, versus a little late, versus very late. Probability theory can work with the probabilities of the different outcomes, but to make decisions, you also need to consider your preferences and your values. Your decision needs to depend on your own personal rating of the desirability or undesirability of the various outcomes.

Can we possibly use cold, austere mathematics to discuss such value-laden concepts as likes and dislikes? Mathematics can never distinguish right from wrong, or good from bad. However, if we can quantify our personal good and bad ratings, mathematics can guide us in making decisions.

To quantify your preferences, you must specify your *utility function,* which gives your personal numerical ratings of all of the different outcomes that may occur. This utility function is positive for good things (the higher, the better), and negative for bad things (the more negative, the worse). For example, you might give a utility rating of +10 for seeing a pretty good movie, or +20 for seeing a really good movie, or +1,000,000 for winning the lottery jackpot. On the other hand, you might give a utility rating of −10 for stubbing your toe, −20 for getting a headache, and −1,000 for getting fired from your job.

Utility functions are a component of *game theory,* the science of decision making, which is often applied to economics, political science, and sociology. These functions were studied in the 1940s by Hungarian mathematician John von Neumann, one of the original six mathematics professors (along with Albert Einstein) at the world-famous Institute for Advanced Study in Princeton, New Jersey. Utility functions provide a simple, clear rule for resolving complicated decisions.

For example, suppose you're busy planning your wedding, and you need to select the venue. You've narrowed it down to two choices: an elegant ballroom in the city or a rustic cabin in the woods. The cabin setting, with its swaying trees and shining lake, is incredibly beautiful, but there is a problem: what if it happens to rain on the day of your wedding?

To deal with this dilemma, you could create a utility function. On a sunny day, a wedding at the cabin would be so delightful that you give it a utility value of +1,000. A wedding in the ballroom (rain or shine) would also be lovely, but not quite at the same level; you give it a utility value of +800.

However, the cabin on a rainy day would be a mess: guests huddled indoors, muddy shoes, a leaky roof, squabbling families, and the beautiful view completely wasted. Your marriage would still be joyous, of course, but your wedding day would correspond to a big fat zero, or a utility value of 0. And, based on past weather patterns, you estimate a 25% chance of rain on your wedding day.

So now your choice is between a dependable ballroom with a value of +800 or a risky cabin that could have a value of +1,000 in the sun, or 0 in the rain. Which should you choose?

At this point, you can begin calculating. The cabin would be worth +1,000 with probability 75%, or worth 0 with probability 25%. This means that if you choose the cabin, the average (or expected) value of your utility function will be 75% of +1,000, plus 25% of 0. This works out to +750. Therefore, if you choose the cabin, your average utility value will be +750. But if you choose the ballroom, your utility value will be +800 regardless of the weather.

Since +800 is more than +750, the ballroom is a better choice than the cabin. So, your (reluctant) choice should be to book the ballroom. That way, your wedding will be a success whether it rains or not. (And, you can always drop by the cabin for a risk-free bonus visit on a sunny day during your honeymoon.) Utility functions have helped you to make a difficult emotional decision using rational, logical thought.

Make that Phone Call

Juan from the finance department seems so nice—and he doesn't wear a wedding ring. Maybe you should invite him to see your friend's rock band on Saturday.

Nervously you reach for the phone. But then you hesitate. What if Juan isn't interested? What if he already has a girlfriend? What if he thinks you're stupid for phoning him? What if he says something mean? You decide the probability that he will accept your invitation is only about 10%. Maybe you shouldn't phone him after all.

Fortunately, you know about utility functions. You decide that if Juan accepted your invitation, the date would be fun and exciting. It might even change your life. You give it a utility value of +1,000.

If Juan declined your invitation, you would be very disappointed, but it wouldn't be so much worse than never phoning at all. Either way, Juan would forever elude you. All that you would really suffer is the embarrassment and nervousness of making the call and expressing your interest. That would be pretty bad, but not too terrible; you give it a value of −50.

So, what is your average utility value from making the call? Well, 10% of the time you would score +1,000, which works out to +100. And 90% of the time you would score −50, which works out to −45. You conclude that making the call has an average utility of +100−45, which equals +55, a net positive.

On average, you stand to gain from making the call. Satisfied but terrified, you pick up the phone. Juan answers, you have a nice chat, you see the rock band together, nature takes its course, and you live happily ever after, all thanks to utility functions.

Insurance Assurance?

Utility functions are also helpful when deciding whether or not to purchase an insurance policy.

When considering insurance, the first questions to ask yourself are, "Over the long run, am I likely to pay more for my insurance than I will receive back in the form of claims payments? Or, am I likely to receive more than I pay?" If you receive back more than you pay in, your insurance policy is a wise investment. But, if you pay more to the insurance company than you receive, your purchase may not be a wise one.

Suppose that your home insurance costs $800 per year. In most years you will not have any home insurance claims at all, and your net gain from your insurance policy will be −$800. On the other hand, every once in a while you might have a major home problem—fire, flooding, burglary, a collapsed roof—and you might recover a claim of many thousands of dollars. Does this small probability of receiving a large amount balance out the $800 policy fee or does it fall short?

It is hard to calculate the answer to this question directly. After all, it depends on such factors as the average frequency of fires or floods; the average amount of damage done by fires or floods; and a calculation of which other events might also lead to an insurance claim. Furthermore, these averages might depend on precisely where you live, the habits of your neighbors, and so on. Nevertheless you can make some educated guesses.

The number one fact about insurance is that insurance companies usually make very large profits. Indeed, insurance is one of the most reliably profitable industries around. But we know from the Law of Large Numbers that the only way a company can make money over the long run is to take in more money than it pays out, on average. Thus, given the profitability of insurance companies, it simply must be true that on average they take in more than they pay out. What this means is that, on average, customers will pay in more than they will receive back.

In other words, without looking up any statistics about fires or damages or insurance prices, we can confidently conclude that, on average, buying insurance is a money-losing proposition. On average, you will pay more money for your insurance policy than you will ever get back in the form of claims payments. Does this mean that no one should ever buy insurance? No, it doesn't, and utility functions tell us why this is so.

Paying $800 per year for insurance is a modest expense to which we can assign a negative utility value of, say, −800. However, if you do not have insurance and experience a major disaster (such as a fire or flood), the consequences may be devastating. For example, if a disaster forces you to sell your home or drives you into permanent financial ruin, it may destroy your life far more than mere dollars can represent. Even if the purely monetary value of your loss is only $100,000, your financial status and security may be gravely disrupted, your marriage may crumble, and your kids may have to leave university, causing you a negative utility of −500,000 or worse. In short, the significance of your difficulties may affect you above and beyond the purely monetary value of the loss.

Suppose that each year there is one chance in 200 of such a major disaster. Then from a purely monetary perspective, you pay $800 for insurance, to have one chance in 200 of collecting $100,000. This means that on average you pay $800, and receive just $500 (or $100,000 divided by 200), for a net loss to you (and corresponding profit to the insurance company) of $300 per year. But your average utility is actually 2,500 (or 500,000 divided by 200) for possibly avoiding a major disaster, minus 800 for paying for insurance, which works out to a positive result of +1,700.

Looked at this way, insurance can sometimes be a win-win situation, with both the insurance company making a profit and the client gaining in utility on average. But this is only possible when insuring against catastrophic losses, for which the perceived value to the client greatly exceeds the purely monetary value of the company's payout.

For non-catastrophic losses, it is always better, on average, to "insure yourself" by *not* buying insurance and paying any losses out of your own pocket. Occasionally you will lose a lot of money, but on average you will save more on insurance costs than you will have to pay out for damages that you incur. In short, you can keep the insurance company's profit margin all for yourself.

Your Utility or Mine?

In one episode of the television sitcom *Happy Days,* Richie and his friends, desperate to meet women, conduct a phony beauty contest. Problems ensue, there is no prize money, the contestants get upset, and chaos reigns. Richie's father is furious, and yells at him, "Was it really worth all this trouble, just to look at some pretty girls?" The episode ends with a close-up of Richie, who is suitably chagrined from his father's scolding, suddenly breaking into a subtle smile. The implication is clear: to adolescent Richie, it *was* worth all the trouble just to look at some pretty girls.

Such conflicts between parents and children arise all the time. We typically dismiss them as a generation gap, or as parents being more mature than their children. In fact, these differences of opinion can be explained by utility functions.

In the above example, Richie was so excited to meet young women that his utility value might have been +100 or more. Meanwhile, although Richie did feel bad about all the trouble he caused, in his youthful exuberance he probably didn't feel *too* bad, so his utility value for the negative fallout might have been −50. Since +100 more than compensates for −50, Richie secretly felt that his joy was worth all the trouble he caused.

On the other hand, Richie's father, if he recognized at all the pleasure of meeting attractive women, might have considered it a minor diversion worth at most +10. But being a responsible citizen in his community, he would have considered all of the resulting trouble and deception and anger to be quite serious, perhaps −100. Since −100 is far more serious than +10, it's no wonder Richie's father was angry.

The Naughty Nephew

Your adorable little nephew is over for a visit, and he is playing with his favorite ball. You told him to be careful, but he is being nothing of the sort. He has already broken a glass, and after you yelled at him he knocked down a picture. How can he be so unreasonable? Obviously, from your point of view, damaging a glass and a picture is far more serious than a bit of fun with a ball.

You then consider your nephew's utility function. He loves playing with the ball, the wilder the better, and might rate the activity at +20. On the other hand, he only feels a little bit bad when he breaks things and gets yelled at, perhaps −10. From his point of view, his actions are completely reasonable.

Part of you feels you should teach your nephew to be responsible by punishing him or taking away his ball. Then again, perhaps his utility function isn't crazy after all. As a compromise, you carefully remove all the remaining breakable objects from the room and allow your nephew to continue playing. Your belongings are safe, your nephew can play wildly with his ball, both of your utility functions are respected, and you are both happy.

Utility functions can also explain the frequent disagreements between doctors and their patients. Often a doctor will recommend a treatment that a patient will not appreciate, even though they both have the patient's best interests at heart. Sometimes the doctor will simply be wrong, or the patient will simply be stubborn and unreasonable. But often the disagreement results from having different utility functions.

For example, suppose a doctor recommends a treatment that reduces your probability of dying by 1%, but at the expense of causing headaches and digestive discomfort—that is, of reducing your quality of life. The doctor feels that the treatment's benefit is worth the discomfort, but you are not sure. Why the difference of opinion?

Doctors are concerned primarily with keeping their patients alive. Your doctor might give a rating of +10,000 to keeping you alive, and 1% of 10,000 is +100. On the other hand, doctors may be less worried about quality-of-life issues (which are harder to measure and quantify and are less frequently studied). Your doctor might give only −20 to the various discomforts that the treatment will cause. Since +100 beats −20, the doctor recommends the treatment.

You probably share your doctor's concern about staying alive, so perhaps you also give a utility value of +100 to a 1% reduction in your probability of dying. On the other hand, quality of life is also very important to you, and you give −200 to the discomfort that the treatment causes. To you the −200 from the discomfort beats out the +100 from increased probability of living, and the treatment doesn't seem like a good idea.

The next time you find yourself in such a situation, don't yell and scream, don't get in a panic, and don't threaten to sue. Simply explain, calmly and politely, "I'm sorry, doctor, but I reject your recommended treatment because I have a different utility function than you do."

7

White Lab Coats

What Studies Do and Don't Show

We are constantly being told just what "studies show." Detergent manufacturers tell us how white our shirts can be if we use their product. Psychologists tell us how to raise our children. Urban planners tell us how to direct traffic. Pharmaceutical corporations instruct us that buying their pills will save our lives. Our doctor tells us that a certain course of treatment is unquestionably the best option. Actors in television commercials, complete with white lab coats and file folders and glasses, inform us of the benefits of sugarless chewing gum. In every case, we are assured that the conclusions are supported by studies. And there's that blabbermouth down at the local bar who claims that "studies show" just about every opinion that he cares to enunciate.

Typically, we trust (or quote or simply make up) studies whose conclusions correspond to our own opinion, but dismiss or ignore those that don't. But is this rational? Can't we learn anything concrete from the studies themselves?

Yes, we can. Indeed, a little understanding and Probability Perspective can teach us which studies to trust, and when.

An Earth-Shattering Conclusion or Just Plain Luck?

A typical medical study might run as follows. A disease (let's call it *Probalitus*) is fatal in 50% of cases. A pharmaceutical company develops a new drug that they claim reduces the fatality level of Probalitus. But does it really?

To test this claim, a study is commissioned. A number of patients suffering from Probalitus are collected together and given the drug. The fraction of fatalities in the study is observed. The question is, how does this fraction compare with the previous 50% probability of dying from the disease?

If the fraction of fatalities in the study is more than 50%, this is a bad sign. The drug is likely a failure. The pharmaceutical company goes back to the drawing board to improve it. In this case, the study has protected humanity from a useless (or perhaps even harmful) drug. So far, so good.

But suppose now that the fraction of fatalities in the study is less than 50%. Suppose, say, that just 40% of the patients in the study died, as compared with 50% who would have died (on average) without this new drug. In this case, the drug sounds promising; perhaps it really does reduce the danger of Probalitus.

Or does it? The question is, does the reduction in fatalities prove that the drug helped, or did we just get lucky? At what point can we accurately conclude that the "study shows" that the drug helped?

Lucky Shot?

"I'm so good at basketball," your boyfriend brags, "that I can score from the other end of the court, nearly every time!"

Tired of your boyfriend's bravado, you decide to put him to the test. You go to the gym together late one night, with a basketball. He stands at one end, bends his knees, and heaves the ball toward the far basket.

Time stands still while the ball sails through the air on a graceful parabola. It's going too far. No, wait, maybe it's not going far enough.

Finally the ball descends, gliding toward the waiting mesh. Down it goes, and then, finally . . . it's in. He did it!

"Yippee," your boyfriend cries. "I *told* you I could do it!"

"Ah, you just got lucky," you retort. "You can't prove anything with one lucky shot. I bet you can't score again."

Your boyfriend sighs. "Aw, come on," he whines. "How many times will I have to score before you believe it's not just luck?"

Good question.

Consider a concrete example. Suppose the study involves just three patients, each of whom suffers from Probalitus. Suppose all three are given the new drug, and all three survive the disease. Bravo, you might think. Without the drug, 50% of patients die, but with the drug, it seems they all survive. Let's get this drug to market right away!

But can we be sure about this conclusion, or did we just get lucky? That is, did the drug actually help the three patients survive? Or did all three patients get better purely by chance, and the drug had no effect whatsoever?

This question is related to another situation, involving our old favorite scenario, flipping coins. (Coins have been used as monetary instruments for about 2,700 years and have undoubtedly been flipped for nearly as long, so it is not surprising that probabilists keep referring to them.) Suppose your friend is dividing up some candies by flipping a coin. For each candy in the box, he flips the coin. If the coin comes up heads, he gets to keep the candy; if it's tails, you get to keep it. Suppose he gets heads on his first three coin flips. Does that mean he is cheating, using a two-headed coin or some other trick? Or, is he being perfectly honest, and he just got lucky? Which is it? How can you tell?

From the point of view of probability theory, the friend counting heads and the Probalitus study counting survivors are exactly the same situation. In both cases, the question is, do the results show an actual result (a beneficial drug or a cheating friend), or did they occur just by luck?

To separate actual conclusions from "just luck," we have to consider what is called the *p-value*. This is the probability that you would have observed such a surprising result (such as three patients all surviving, or getting three heads in a row) just by pure luck, if the drug had no effect and the friend didn't cheat.

In the Probalitus study, if the drug had no effect, each patient would have a 50% chance of surviving. So, the probability that all three patients survive would be 50% of 50% of 50%, or 12.5%. In reporting the Probalitus study, we could say that the drug helps reduce the disease's fatalities, with a p-value of 12.5%.

Similarly, the probability of getting three heads in a row by just luck is also 12.5%. In either case, we can say that the p-value of the study is equal to 12.5%.

What is the meaning of p-values? Well, if the p-value is large, then maybe you just got lucky, and the study proves nothing. But if the p-value is very small, it is highly unlikely that your results were purely the result of luck, and therefore quite likely that your drug had a positive impact.

In the Probalitus study, the p-value is 12.5%. Is 12.5% a small enough p-value that we should recommend this drug to all who suffer from Probalitus? Or is 12.5% large enough that we can dismiss the results of this study as being from just luck, and therefore of no significance?

How Unlikely Is Too Unlikely?

We should always be careful not to jump to premature conclusions, and to avoid doing so, we require a standard for how small a p-value must be before a study's result can be considered to represent a true finding. In medical and psychological testing and elsewhere, the traditional standard is 5%. That is, any study result that has less than a 5% p-value, or less than one chance in 20 of happening just by luck, is considered to be "statistically significant." On the other hand, if the p-value is more than 5%, the results might be due to pure chance, and they are therefore not statistically significant.

The Suspicious Subway Stroller

You're riding the subway, and something catches your eye. It's a large man who is a bit unusual looking. What is it that bothers you so?

That mustache. Red and bushy, just like the KGB agents in all those spy movies. That's it, this guy is a leftover KGB spy!

Calm down, you think. It's possible that a regular, ordinary guy might have a bushy red mustache. That doesn't prove a thing.

But those boots. Steel-toed, black, with a thick heel. Surely only a KGB agent would wear such boots!

No, no, you remind yourself. It's conceivable that an honest, law-abiding citizen might wear heavy, black steel-toed boots.

Hold on, what about that bulge in his overcoat. It must be a gun. He must be a renegade KGB agent, trying to undermine your way of life! You have to do something. There is no time to lose.

To keep the world safe for democracy, you leap forward and tackle the man. Stunned and surprised, he does not resist.

The police arrive, and a careful check reveals that this man is an honest wheat farmer from Saskatchewan who just happens to have a bushy red mustache, just happens to like heavy black boots, and just happens to keep his wallet stashed in his coat pocket. He is released, with full apologies.

You, on the other hand, spend the next six months in prison after being convicted of unprovoked assault.

What about our Probalitus study? In that case, with just three surviving patients, the p-value is 12.5%, which is much more than 5%. Thus, this study does *not* in any way establish that the drug actually reduced the fatality of the disease.

On the other hand, if our Probalitus study had *five* patients, and they all survived, the p-value would equal 50% multiplied by itself five times, or 3.1%. This is less than 5% and is thus statistically significant. We could then conclude that the result wasn't just luck, and that the drug had actually helped to combat the disease.

Similarly, if your friend gets three heads in a row, he might just have gotten lucky, and you should give him a break. But if the pattern continues and he gets five heads in a row, you might want to carefully examine the coin that he's flipping.

The 5% maximum error probability is very widely used throughout the sciences. It is viewed as the statistical equivalent of the lawyer's standard of "beyond a reasonable doubt." However, the precise figure of 5% is actually quite arbitrary. Furthermore, it allows for the possibility that as many as one medical study in 20 may be wrong. The 5% standard was first chosen in the 1920s by R.A. Fisher, a British agricultural researcher and a father of modern statistical inference, who found the figure to be both appropriately small and mathematically convenient. Some statisticians feel that, to avoid false conclusions, statistical significance should require the p-value to be less than 1%. (To achieve a p-value less than 1% would require observing seven surviving Probalitus patients—or seven coins coming up heads in a row—not just five.) Even Fisher allowed that "if one in twenty does not seem high enough odds, we may, if we prefer it, draw the line at one in fifty, or one in a hundred."

The debate continues, but in the meantime, 5% remains the standard in almost all fields of academic research. Fortunately, many published studies report the actual p-value obtained, so if you want you can check it yourself. In some cases, the p-value might be 4.9%, while in other cases it might be very small (say, 0.5% or less). The study's p-value is an indication of how confident you can be in the conclusions reached: the smaller the p-value, the less probability that the result was really just luck. So, the next time your doctor quotes a medical study, don't accept the conclusion blindly. First, enquire about the study's p-value.

Remembering that events sometimes occur "just by luck" can also help us understand some everyday events. The principle of *regression to the mean* says that extreme or surprising observations are often caused by a combination of true differences and plain old luck, so they are likely to lessen (i.e., return to more "usual" values) in the future. For example, if a student does extremely well on a test, she is probably quite smart but she probably also benefited from a bit of luck. On the second test, she is likely to do pretty well, but not as well as she did on the first test—her luck may have run out. Meanwhile, a student who did extremely poorly the first time around might well improve a little on the second test. For another example, if a man is extremely tall, then his son is most likely to be taller than average, but less tall than his father. Or if an athlete has an outstanding game, probably the next game won't be quite as good. Regression to the mean is sometimes used by investors who reason that if a stock's price suddenly falls, perhaps it got unlucky and will soon rebound, so now may be a good time to buy some more shares. (Of course, there is no guarantee; perhaps there was a reason for the stock's decline, which will only get worse.) Regression to the mean is a statistical version of what all parents tells their children after a really rotten day: tomorrow will be better. (On the flip side, if today was a really great day, then tomorrow might be a little worse.)

Beware the Bias

Our rule, now, is quite clear. To test the validity of a claim (that a drug improves health, or a detergent works better, or anything else), we conduct a study, and compute the p-value that the result was just luck. If the p-value is less than 5% (or, even better, 1%), then the study's results are statistically significant, and we can trust the study's conclusion.

So far so good. However, for a study to be valid, we must ensure that it is not biased. There are several types of bias to worry about. One occurs when patients are selected for study based on their condition.

Army Inspection

The drill sergeant greets you with a smile and a handshake. "Hello, Colonel. I think you'll find that my troops all approve of the way I lead them. Let's go and ask them."

He leads you into the quarters in which his men are assembled. "Bender," he calls to one of them, "what do you think of my troop leadership?"

Bender hesitates. "Uh, well, sir," he begins cautiously, "you do have a tendency to—"

"Silence, Bender!" the sergeant screams. "Calder! What do you think of my troop leadership?"

Calder stands up and begins quietly, "Well, to be perfectly frank, sir, it is my opinion that—"

"Shut up, Calder!" the sergeant growls. "Fawler! What do you think of my troop leadership?"

Fawler gets to his feet. "Why, I like it, sir. I think your tough approach is just what we need."

"Thank you, Fawler," the sergeant replies.

Turning to you, he continues, "See, Colonel, I told you they all approve of my leadership!"

If the manufacturers of the anti-Probalitus drug are sufficiently desperate to prove their drug's value, they could resort to a trick. They could conduct a study in which they give their drug only to patients who seem to be recovering anyway. That way, most of the patients in the study will recover—not because of their drug, but because they were doing well in any case. The p-value for this study would indeed be very low, but only because they cheated.

This problem is called *sampling bias*. To avoid it, studies should (and usually do) assign their patients randomly. For example, for each Probalitus patient who volunteers to participate in the study they might flip a coin and give the drug only if the coin comes up heads. That way, every patient has the same probability of getting the drug or of not getting the drug, regardless of their level of health.

We've all seen testimonials on television from famous musicians and actors about how they came from humble roots, and struggled without pay for many years, before finally being discovered and striking it rich.

Many aspiring entertainers are encouraged by such stories to devote all of their energies to their craft, with expectations of similar glory. The reality, however, is that tens of thousands of people aspire to be rock stars and never succeed. It is only those few who do succeed who are interviewed on television. The resulting interviews form a biased sample of struggling musicians and give a very misleading picture.

The media's tendency to over-hype murders, discussed earlier, can also be viewed as a form of bias. Each day, about 99.99998% of people are not murdered, yet the media focus on those few who are. This may be reasonable from a news perspective, but it results in newspaper headlines that give a very biased view of the society they attempt to describe.

A similar problem arises from *reporting bias*. For example, perhaps when totalling up the study, a company conveniently "forgets" to report some of the fatalities among patients taking their drug. Or, perhaps they conveniently determine that some of the patients have recovered before they really have. Any such discrepancies skew the results of the study, thereby rendering them invalid.

To avoid such biases, studies should not be conducted by people or companies with a vested interest in the outcome. Rather, a disinterested, objective, independent professional organization should conduct the study and report the results, without playing any favorites. That way, the sampling can be done fairly, and the reporting can be done accurately, without in any way influencing the results.

Biases can be subtle and they can explain mysterious practices. For example, in many jurisdictions, adult citizens are randomly selected for jury duty. Once such citizens have served, they are excused from further service for a period of a few years. But why are they not excused forever, or at least until everyone else has already served at least once? The answer is that such a rule would create bias. Jury selections would gradually become weighted toward the few citizens who have not already served, namely young adults and new immigrants. The powers that be have decided that such juries would not be appropriately representative of the entire population. This is why the exclusions do not last forever, and why some people serve three or four times while others never do. It may not seem fair, but at least it avoids bias.

Reporting bias sometimes takes surprising forms. A student I know was recently phoned for a survey, conducted by a university student union, about student debt loads. He explained that he does not have any student debt as such, but that he has a large mortgage on a condominium he recently purchased; should he count that debt or not? The surveyor thought for a moment about this unexpected development. He then stated that counting this debt would help their cause (which was to illustrate the problem of large student debt), and so he counted it. Clearly, if such decisions are made in the context of what result a study's sponsors might want, the entire study is rendered invalid.

The Discouraging Diet

You want to lose some weight, so you read about various popular diet plans and resolve to follow them. For breakfast you select steak and eggs with butter and bacon, content with your low-carb diet. For lunch you devour large portions of white bread with honey, a soft drink, and two Popsicles, happily following a low-fat diet. For dinner you eat three slices of whole-grain bread loaded with peanut butter, a large plate of whole-wheat pasta with tomato sauce, and a big scoop of brown rice, proud of your high-fiber diet. For an evening snack you have one scoop of ice cream with chocolate topping, thus carefully following a diet based on small portions.

The next day you are shocked to discover that your weight is up again, after you were so careful to follow a diet every time you ate. Only then does it occur to you that your self-assessment as a careful dieter might be biased, since you have been choosing which diet to follow based on what you feel like eating at each moment. From now on, you had better pick just one diet and stick to it, thus avoiding bias and perhaps even losing some weight.

A similar problem arises with movie advertisements that quote reviewers. Virtually every movie is liked by *some* reviewers, and excerpts from those reviews inevitably appear in advertisements for the movie. Movie watchers with Probability Perspective ignore such ads; they instead pick one or a few specific reviewers (in my case, it's usually Roger Ebert), and base their movie choices on what those reviewers say. It is not that those reviewers are necessarily wiser or more insightful than all of the others, but sticking with the same reviewers when comparing different film ads avoids the bias of movie promoters who select which reviews they use to publicize their product.

Publication Bias

Now we know that studies should be conducted by independent professionals, and that the results should be interpreted in terms of p-values. So that's that, right?

Not necessarily.

As children, we learned the trick that if your mom says no, then you ask your dad, thus nearly doubling your chances of being allowed out to play. Unfortunately, there is a simple way for large companies to do the same thing, called "publication bias." This involves commissioning many studies and only publishing those whose conclusions are favorable to the company in question, while burying the rest.

The Happiness Hat Hoax

As the CEO of Happiness Hats, Inc., you are getting concerned. Your company is very successful, but lately sales are falling. How can you regain the public's interest?

You decide to commission a study about the benefits of Happiness Hats. You hire a leading expert to conduct the study. The expert gathers together a few hundred subjects and randomly divides them into two groups—one wearing Happiness Hats, the other not. One month later, the happiness levels of the two groups are measured and compared.

At the end of the study, the expert gives you a full report, which you eagerly read. But your enthusiasm soon turns to despair. The study's conclusion is that wearing Happiness Hats made *absolutely no difference* to the happiness level of the subjects.

You're not a quitter, so you try again. This time, you employ 100 different independent experts. You pay each of them to conduct a fresh Happiness Hat study. For good measure, you put a little clause in their contract that says that their final reports should be given only to you and shared with no one else.

Finally the reports come in. Unfortunately, most of them again conclude that Happiness Hats make no difference. In fact, a few of them even conclude that Happiness Hats are harmful. However, in study #57 (oh, glorious #57!), the Happiness Hat wearers happened to have a run of good luck. The study therefore came to the conclusion that Happiness Hats increase happiness. The p-value was nice and small, well under 5%. Yippee!

You publicize study #57 widely and broadly. You publish it in journals. You advertise it on television. You quote it on the radio. You plaster it on billboards throughout the land. When skeptics enquire further, you introduce them to the expert who conducted study #57; he confirms that he was totally unbiased and that he performed his study carefully and fairly. Everyone is duly impressed, Happiness Hats enjoy a huge resurgence, and your corporation gets richer than ever.

What about the other 99 studies? Well, let's just say that your high-capacity paper shredder got a lot of use that night.

You might think that no company would ever hide or shred studies that they had commissioned. In fact, medical-research funding contracts from pharmaceutical companies (and others) often stipulate that findings may not be published without the company's consent, so the company can control which studies are publicized and which are not.

In 1996, Dr. Nancy Olivieri, a researcher at Toronto's Hospital for Sick Children and an acknowledged world expert on internal medicine and hematology, was commissioned by the pharmaceutical company Apotex to conduct clinical trials using Apotex's drug deferiprone to treat children with the blood disease thalassemia. After some time, Dr. Olivieri became convinced that deferiprone sometimes caused liver fibrosis, a potentially fatal form of scarring of the liver. She felt that the problem was serious enough that it should be brought to public attention. However, the contract that Dr. Olivieri had signed with Apotex prohibited publication of her results within three years without Apotex's explicit consent (which they withheld).

Dr. Olivieri decided to go public anyway. Her findings were published in the *New England Journal of Medicine* in 1998. In response, Apotex threatened legal action, and Dr. Olivieri's research position at the Hospital for Sick Children was revoked (although later she was reinstated at the neighboring Toronto Hospital). The situation caused a huge controversy throughout the medical research community. Much of the controversy was caused by the realization that while the Olivieri case was the most high-profile incident of medical research being influenced by funding agencies, it was far from the only one.

Some of the ensuing debate focused on the validity of Dr. Olivieri's findings: did deferiprone cause harm or not? Some experts claimed it did, while others argued that it did not. From a medical perspective, it is of course important to evaluate accurately the risks of deferiprone (and of all other drugs). But from a Probability Perspective, what is at issue is publication bias. The decision to publish research results should be based only on the quality of the research itself, not on who may like or dislike its conclusions. Such censoring serves to bias the results away from the objective truth, and thus renders p-values and other statistical conclusions meaningless.

There have been other, similar examples of publication bias. In 1990, Dr. Betty Dong, a clinical pharmacist at the University of California at San Francisco, determined that the thyroid medication Synthroid was no more effective than cheaper alternatives. The manufacturer of Synthroid, the Boots Company, was furious. They tried to persuade Dr. Dong to modify her conclusions, and when she refused they hired consultants to discredit her study. When Dr. Dong's study was about to be published in the *Journal of the American Medical Association* in 1994, Boots took legal action, citing the study's funding agreement, which said that publication of results was forbidden without Boots's consent. The study was finally published in *JAMA* in 1997 but only after extensive publicity and many high-level meetings between company and university officials. (By then Boots had been acquired by another, apparently less confrontational, company.)

The recent controversy involving the anti-inflammatory drug Vioxx can also be viewed as a form of publication bias. In September 2004, this drug was withdrawn by its manufacturer, Merck & Co., due to accumulating evidence that it significantly increased the risk of heart attacks. Indeed, a study published in *The Lancet* in January 2005 estimated that Vioxx was probably the cause of between 88,000 and 140,000 cases of serious coronary heart disease. It was reported by *The Wall Street Journal* and others that researchers at Merck had suspected this risk years earlier. Internal company e-mails as far back as 1997 noted that "the possibility of increased CV [cardiovascular] events is of great concern," and by March 2000 Merck had acknowledged internally that the cardiovascular problems associated with Vioxx "are clearly there." However, the company kept this information to themselves and continued to assert that Vioxx was safe, issuing a news release in April 2000 headlined "Merck Confirms Favorable Cardiovascular Safety Profile of Vioxx." One internal training document even instructed Merck marketers to "dodge" questions about the cardiovascular effects of Vioxx. Additionally, in 2002 Merck sued a Spanish medical researcher, Dr. Joan-Ramon Laporte, who was critical of Vioxx and of Merck (the suit was eventually thrown out of court). Professor James Fries of Stanford University Medical School wrote that several medical researchers who were critical of Vioxx had been phoned and threatened in "a consistent pattern of intimidation of investigators by Merck." While the truth about Vioxx's dangers eventually became public, it took years longer than it should have. We should all worry about which other drugs' truths still remain hidden.

These and other controversies have put a spotlight on the widespread practice whereby corporate funders of medical research, who often have a vested interest in the results, have the final say over which studies get published. This practice creates the risk that companies will publish only those studies that are favorable to their products. If so, the p-values and other probabilities associated with each individual study are entirely misleading, since they do not take into account the bigger picture of selectively published data. As anti-corporate crusader Ralph Nader might have put it, "Unsafe at any p-value."

The medical community is finally taking this issue seriously. The International Committee of Medical Journal Editors—a group of editors of 11 leading medical research journals including the *Journal of the American Medical Association, New England Journal of Medicine, The Lancet,* and *Canadian Medical Association Journal*—declared in 2001 that from now on, in all studies published in their journals, "the sponsor must impose no impediment, direct or indirect, on the publication of the study's full results, including data perceived to be detrimental to the product."

Clearly, those editors were using Probability Perspective.

Cause and Effect

Even if studies are conducted properly and objectively, with accurate use of p-values and full freedom to publish the results, we must still be careful to interpret the results correctly. Often, results that seem to imply one thing actually imply another.

The Jumping Frog (An Old Joke)

You want to determine the relationship between a frog's legs and its ability to jump. So, you find a frog and a measuring stick, place the frog at one end, and say, "Jump!"

Dutifully, the cooperative frog leaps into the air and lands at the 82-cm mark. "Aha," you declare, "frogs with all four legs can jump 82 centimeters."

To continue the study, you get out a sharp scalpel and slice off the frog's left front leg. (The frog, always eager to push the frontiers of science, remains stoic.) You return the frog to the starting point. "Jump!" you call again.

The frog leaps into the air and this time lands at the 47-cm mark. "Aha," you say excitedly, "frogs with three legs can jump 47 centimeters."

Slicing off the frog's right front leg, you repeat the experiment again, with the frog landing this time at the 18-cm mark. "Aha," you marvel, "frogs with two legs can jump just 18 centimeters."

You then slice off the frog's left back leg. The frog slithers awkwardly and lands at the 5-cm mark. "Wow," you exclaim, "frogs with one leg can still jump five centimeters."

Finally you slice off the frog's remaining leg. "Jump!" you order, but the frog does not move.

Annoyed, you repeat, "Jump!" Again there is no reaction. "JUMP!" you shout, but the frog remains motionless.

"This is fascinating," you gush. "What an interesting conclusion."

Picturing yourself accepting next year's Nobel Prize, you practice announcing your astounding finding: "Frogs with no legs are deaf!"

To better understand the difficulties of determining causality, consider a classic (albeit exaggerated) example. It is now well established that smoking cigarettes significantly increases the risk of contracting lung cancer, but this connection was once controversial. It also happens to be true that cigarettes sometimes cause mild (and harmless) yellow stains to appear on smokers' fingers.

Consider a researcher who didn't know much about smoking, but who wanted to determine what causes lung cancer. She might notice that many of the people with lung cancer also had yellow stains on their fingers. Of course, this would *really* be because both factors are themselves caused by smoking. But if the researcher didn't know that, she might erroneously conclude that *yellow finger stains cause lung cancer*.

Imagine the difficulties that such a false conclusion could cause. Parents would refuse to let their children play with yellow crayons, lest they stain their fingers and get cancer. Smokers might don latex gloves, so that they could smoke "safely" while avoiding those harmful yellow stains, all the while filling their lungs with tar and nicotine that they don't know could kill them. Family doctors might turn to scrutinizing patients' fingertips under a microscope and neglect such trivialities as listening to their wheezing lungs. A simple misunderstanding of what causes what could have grave consequences.

Statisticians have a saying to deal with this problem: "Correlation does not imply causation." Just because two traits (like getting lung cancer and having yellow finger stains) tend to go together (i.e., are *correlated*), that does not necessarily prove that one causes the other. While no one would seriously conclude that yellow stains cause cancer, the issue of what causes what arises often and in many contexts.

The Meditation Medical Miracle

The data seem incontrovertible: Participants in the Million-Minute Meditation (MMM) program are far healthier than the general public. For a modest $10,000 annual fee, participants spend two hours a day engaged in concentrated meditation and spiritual awakening, under the direction of the experienced meditator Wily Wally. Careful medical examination has shown that participants in MMM have lower blood pressure, less body fat, greater muscle strength, lower cholesterol, and greater lung capacity than the population as a whole.

Wily Wally is trying to convince you to join the MMM. "Don't you care about your health?" he asks. "Wouldn't you like to be as healthy as the other MMM participants?" Furthermore, for a limited time only, he agrees to knock $10 off of the annual fee.

Before you plunk down the money, your Probability Perspective kicks in. It is indeed possible that all this meditation is responsible for impressive health benefits. However, it is equally possible that the MMM participants are a self-selected group. That is, anyone who is willing to pay $10,000 a year and put in two hours a day to improve their health must care about their well-being a great deal, indeed. Probably they also exercise regularly, eat well, avoid stress, get regular medical checkups, and otherwise take good care of themselves.

If so, rather than the MMM program causing good health, it may be that healthy habits cause people to join MMM, and that MMM has little or no direct impact on health improvement. Thus, rather than join MMM, you might be better off simply exercising regularly and eating well.

"Correlation does not imply causation," you snap at Wily Wally and choose to spend your money elsewhere.

Once we recognize the false-cause issue, we see it everywhere. For example, a recent long-term study of University of Toronto medical students concluded that medical-school class presidents lived an average of 2.4 years less than other medical-school graduates. At first glance, this seemed to imply that being a medical-school class president is bad for you. Does this mean that you should avoid being medical-school class president at all costs?

Probably not. Just because being class president is correlated with shorter life expectancy does not mean that it *causes* shorter life expectancy. In fact, it seems likely that the sort of person who becomes medical-school class president is, on average, extremely hard-working, serious, and ambitious. Perhaps this extra stress and the corresponding lack of social and relaxation time—rather than being class president per se—contribute to lower life expectancy. If so, the real lesson of the study is that we should all relax a little and not let our work take over our lives.

Or there may be other explanations. One somewhat outlandish one is that people who come from families where many people died young are more likely themselves to die young but are also more likely to try to make the most out of what time they have available by doing things like becoming medical-school class president. The point is, based on this study alone, we just don't know why these medical students die younger than their peers. We can be reasonably confident (and surprised) that people who become medical-school class president have shorter lives on average (this is the correlation part). However, we cannot be certain of the causation part: whether being class president causes an earlier death, or whether fearing earlier death causes one to become class president, or whether being extremely ambitious causes both, or what.

There are now a number of studies that indicate that people who watch more television are also, on average, more likely to commit violent crimes. Aha, we might think, this proves once and for all that television is evil, and that all the desensitizing violence on TV causes people to become more violent. But does it? Perhaps people who come from disadvantaged backgrounds or dysfunctional families tend on average to be more violent (due to increased desperation and less parental supervision); they tend also to watch more TV because they lack money for other entertainments or are exposed to fewer other, more rewarding pursuits. The studies convincingly show a *correlation* between violence and television, but they do not establish the underlying *causes* of the behavior. Once again, correlation does not imply causation.

Randomized Trials

If many studies fail to tell us accurately what causes what, do we have any hope of ever getting reliable information?

Fortunately, the answer is yes. The key is to use a randomized trial. What this means is that a study's subjects are assigned at random to one of two different groups, without regard to any other factors about their health or wealth or anything else. The two groups are then given different treatments, such as drug versus no drug, or MMM membership versus no membership. Then, if there is a statistically significant difference in the results between the two groups, it cannot be due to any other factors. Instead, the change has to have been caused by the actual difference in the treatment between the two groups.

We have already seen that in the Probalitus study, the best method is to flip a coin for each patient and give the drug only to those whose flip comes up heads. The patients are thus divided randomly into two groups, a treatment group that gets the drug, and a control group that does not get the drug. If the patients in the treatment group have a much better survival rate than the patients in the control group, then it must be true that the drug is making them better.

Consider again the question of lung cancer and yellow finger stains. Suppose that rather than simply observe who has yellow fingers and who has lung cancer, we take a more active role. Specifically, we do a study by flipping a coin for every patient and painting their fingers yellow if the coin comes up heads (while leaving their fingers alone if the coin is tails), regardless of whether or not they smoke.

In that case, we would (probably) find that there wasn't much difference between the rates of lung cancer for the two different groups. We would then conclude, correctly, that yellow fingers do *not* cause lung cancer after all. They are indeed correlated with lung cancer, but they do not cause it.

Could we then do a similar study to determine whether smoking causes lung cancer? Yes, but it would be more difficult. We would have to flip a coin for every patient and then force them to smoke for many years if the coin came up heads, while not allowing them to smoke at all if the coin came up tails. Naturally, it would be difficult to get patients to cooperate in such a study. Instead, more indirect methods must be used.

A similar problem arises when studying the relationship between television and violence. To conduct a true randomized trial, we would have to flip a coin and force half of the subjects to watch lots of television (every day, for many years), and the other half to watch almost none. If the randomly chosen television watchers committed more violent crimes, this would be conclusive proof that watching television causes violent behavior, and that restrictions on television content might be called for. However, it is hard to imagine how such an experiment could be performed in anything resembling a free society.

These imaginary scenarios illustrate why it is difficult to do proper randomized trials for studies of "lifestyle" factors (like smoking, eating,

TV viewing). However, in medical studies, researchers can control precisely what drugs are provided to each patient, so this is not a problem.

There is one additional factor with medical studies. Patients' health might sometimes improve not because of the drug's actual effects, but because the patient *believes* that the drug will help, and this very belief is enough to make them feel better. To avoid this problem, most medical studies involve giving a placebo (fake pill) to the patients in the non-drug group. That way, the patients don't actually know whether they are getting the new drug or not, thus eliminating any related psychological factors. (In fact, most medical studies are done "double blind," so that even the doctor does not know until later which patients received the drug and which patients received the placebo pills, thus avoiding even subtle cues from the doctor.)

Randomized drug trials can sometimes lead to ethical issues. For example, experts believe that an antiretroviral drug regimen helps slow the spread of HIV/AIDS. Thus, ethical considerations might dictate that all HIV/AIDS patients be given the most promising antiretroviral drugs. On the other hand, to get scientific evidence about the effectiveness of these drugs, researchers need a control group of patients who are not getting the drug.

This conflict between the short-term ethical action (giving the best drugs to all patients) and the long-term scientific action (doing a controlled study where some patients do not get drugs) is difficult to resolve. My own reaction to such controversial questions is to feel glad that I am not a medical researcher.

8

Ain't Gonna Happen

Very Low Probabilities

It is a truism that from time to time unlikely things happen. They may strike us as strange and surprising coincidences, though many of those can be explained by considering "out of how many." They may be events that we have previously dismissed as impossible, using the "ignore the extremely improbable" philosophy. In any case, events of very low probability are intriguing in a variety of ways.

When people think of unlikely events, being struck by lightning comes to mind. We have all seen lightning flash and heard thunder crash during many different storms. Still, it is very rare that lightning actually strikes people, and even more rare that it kills them.

Just how rare? We already saw that there were just 50 deaths from lightning in the United States in the year 2000. Indeed, this is typical: the National Lightning Safety Institute reports 756 deaths from lightning in the United States during the 14-year period from 1990 to 2003, for an average of 54 lightning deaths per year. Compared with the total of about 2.5 million deaths per year, only one person in 50,000 will ever die from lightning. In a single year, only about one in 6 million Americans will be killed by lightning. That's rare.

The infrequency of lightning deaths was put to good use in a recent antismoking commercial, which showed a woman standing at the top of a hill during a lightning storm, holding a tall metal staff, just daring the lightning to strike. The woman explained that her actions may seem crazy, but they were nothing compared with the stupidity of smoking. Since we have seen that lung cancer accounts for about 7% of all deaths, which is about 3,000 times as many as are caused by lightning, the producers of that commercial clearly had Probability Perspective.

On the other hand, lightning is not quite an equal-opportunity killer. You are more likely to be killed by lightning if you live in an area with many electrical storms, if you are often outdoors during storms, and if you live in a flat area without many tall buildings to absorb lightning. During that 1990 to 2003 period in the United States, there were 126 lightning fatalities in Florida and 52 in Texas, but only two in Massachusetts. Table 8.1 shows the most dangerous states for lightning fatalities, 1990–2003, based on the average annual rate per 100,000 people.

Table 8.1 Most Dangerous States for Lightning Fatalities, 1990–2003

State	Population (2000)	Total #	Annual Rate
Wyoming	493,782	14	0.203
Utah	2,233,169	22	0.070
Colorado	4,301,261	39	0.065
Florida	15,982,378	126	0.056
Montana	902,195	7	0.055
New Mexico	1,819,046	14	0.055

By contrast, heavily populated Massachusetts and California are much safer when it comes to lightning.

Table 8.2 Safest States for Lightning Fatalities, 1990–2003

State	Population (2000)	Total #	Annual Rate
Massachusetts	6,349,097	2	0.0023
California	33,871,648	8	0.0017
Alaska	626,932	0	0.0
Hawaii	1,211,537	0	0.0
Rhode Island	1,048,319	0	0.0

What about other countries? The Web site nationmaster.com reports that the largest number of annual lightning deaths happens in Mexico (223), Thailand (171), South Africa (150), and Brazil (132). However, as a rate per 100,000 people, the most dangerous countries for lightning are as follows:

Table 8.3 Countries with Highest Annual Lightning Fatality Rates

Country	Population	# Fatalities	Rate per 100,000
Cuba	11,263,429	70	0.621
Panama	2,960,784	17	0.574
Barbados	277,264	1	0.361
South Africa	42,768,678	150	0.351

Of course, even for these countries, deaths by lightning represent just a tiny fraction of the total number of deaths each year. Notice how poor Barbados, with just one lightning fatality, got ranked as the third most dangerous country simply because its population is so small that even one fatality is a lot.

The uneven distribution of lightning can help to explain some other coincidences. For example, in 2002, Jorge Marquez was struck by lightning for an incredible *fifth* time in his life. (He survived uninjured, though the first time his hair was burned and he lost his fillings.) Given that lightning strikes are so rare, Marquez's experience seems to defy all probabilities. On the other hand, Marquez is a Cuban farm-worker, and thus presumably spends a lot of his time outdoors, even in the rain. Plus, we have just seen that Cuba is one of the most dangerous countries for lightning strikes. So, Marquez is more likely to be struck by lightning than is a randomly chosen person. Even so, it must be said that he is a rather unlucky fellow.

Speaking of bad luck, assistant director Jan Michelini was struck by lightning twice while in Italy filming Mel Gibson's movie *The Passion of the Christ* (though he wasn't seriously hurt). It is unknown whether or not divine intervention was involved.

Unlikely or Not

Some events are so unlikely that they have become clichés. What parent hasn't told a child that he has to clean his room unless he wins the lottery, or that he's more likely to be struck by lightning than to get an increase in his allowance? (The phrase "until hell freezes over" also comes to mind, but I lack the data to compute the probabilities for that one.) Usually such references are harmless and amusing, even if not very accurate, but one such comparison became a part of popular culture in a rather harmful way.

In the mid 1980s, researchers at Yale and Harvard began a study about Americans' marriage patterns and their relation to the age structure of the country's population. They made a preliminary conclusion that a 40-year-old unmarried woman had just a 1.3% probability of ever getting married. This figure was mentioned in a small 1986 Valentine's Day article in the *Advocate* newspaper, published in Stamford, Connecticut. The story was then picked up by major news organizations throughout the United States and beyond, leading to widespread talk of a "marriage crisis" among baby-boom women. The topic dovetailed with the anti-feminist conservatism of the time, and it was implied that women's economic equality was to blame for this "man shortage," because men preferred to marry women of lower economic status. *Newsweek* wrote a cover story that said, "For years bright young women single-mindedly pursued their careers, assuming that when it was time for a husband they could pencil one in. They were wrong."

And then, in an affront to probabilists everywhere, *Newsweek* declared that a 40-year-old single woman was more likely to be killed by a terrorist than to marry. This "fact" struck a chord and was repeated in popular media and around the water cooler for years. In addition to being hurtful toward women, causing unnecessary panic, and implying that marriage was every woman's number-one goal, *Newsweek*'s claim was completely false.

We know that even in 2001, only about one American in 94,000 was killed by terrorists, which is equivalent to 0.001%. This is a much, much smaller figure than even the 1.3% probability of marriage that the study claims. (And in the years before 2001, when terrorist actions on American soil were practically nonexistent, the corresponding probability would have been even smaller.) So, there was simply no logic nor accuracy whatsoever to *Newsweek*'s claim. (A *Newsweek* bureau intern later asserted that the terrorist reference had started out as an office joke.)

It wasn't long afterward that even the figure of 1.3% was questioned. As detailed by Susan Faludi in her book *Backlash*, demographer Jeanne Moorman of the U.S. Census Bureau studied the 1980 census data directly and determined that the probability of a 40-year-old single woman getting married was between 17% and 23%. Her colleague, a statistician named Robert Fay, then reexamined the original, unpublished Harvard–Yale study, which had been based on a questionable statistical model and had assumed that women would necessarily marry men who were a few years their senior. Fay found other errors in the study, and when he corrected for them, he obtained results similar to Moorman's. Fay wrote to the study's authors that he believed that "this reanalysis points up not only the incorrectness of your results, but also a necessity to return to the rest of the data to examine your assumptions more closely." Despite the huge publicity surrounding the original study, these corrections and clarifications were mostly ignored.

What is the probability that a never-married 40-year-old woman will eventually marry? This question involves projecting into the future, so there are a variety of conflicting statistical approaches that could be taken, and every answer is open to some debate. Furthermore, social customs and perceptions around aging keep changing, and women who are 40 today may have different life patterns from those who were 40 in the past. Despite these difficulties, it is possible to attempt to compute a probability by following the marital status of a fixed group of women as they age over a period of many years.

For example, the U.S. Census Bureau reported that in 1970, of all American women between 40 and 44 years old, 4.9% had never married. In 2001, it reported that of the 8,851,000 American women who were over 75 years old, 366,000 (4.1%) had never married. A woman who was 40 in 1970 turned 70 in the year 2000, so these two statistics refer to essentially the same group of women. (Of course, they are not precisely the same group, due to immigration, emigration, and death. But they constitute the same "cohort" of women.) By this reasoning, we could say that the fraction of this cohort who were unmarried at 40 but who were wed by age 75 is about 0.8% divided by 4.9%, which is equal to 16.3%. In other words, of those American women in the year 1970 who were between 40 and 44 years of age and who had never married, about 16.3% eventually married. This is quite a large percentage, especially if you consider that many of the women might not have *wanted* to get married. Furthermore, this figure is over 12 times larger than that claimed by the Harvard–Yale study, and over 15,000 times larger than the chance of being killed by terrorists, even during the 9/11 attacks. And, a 40-year-old woman today is probably even *more* likely to marry than was her 1970 counterpart.

While it is impossible to compute with certainty the probability that a never-married 40-year-old woman will eventually marry, figures such as the 17% to 23% reported by Moorman, or the 16.3% computed here, seem approximately correct. In any case, it is now widely accepted that the original Harvard–Yale study was flawed, that its figure of 1.3% was entirely incorrect, and that *Newsweek*'s analogy to being killed by terrorists was way off base.

Little Green Men?

One of the greatest "coincidences" of all is that life exists on this planet. For humans to have evolved here required, first of all, the formation of an appropriate star (the Sun). It then required the formation of an appropriate planet (the Earth), having the right amount of water and air and land, at a reasonable temperature. Furthermore, life then had to be created and to evolve—over billions of years—into a species as intelligent as we are.

Everyone agrees that the emergence of intelligent life was extremely unlikely. But the question is, just *how* unlikely?

In some ways this is a moot point. After all, we are here, no matter how unlikely that may be. However, the probability that we evolved here is closely related to the probability that intelligent life has also evolved somewhere else.

The discovery of intelligent life elsewhere in the universe, if it ever happened, would be a momentous event. It would directly affect our safety (if the aliens were hostile), our knowledge (if the aliens taught us), our technology (if the aliens let us examine their machines), and our philosophy about our place and importance in the universe. Life would never be the same again. But will such a discovery ever happen?

The organization SETI (Search for Extraterrestrial Intelligence) has been preoccupied with precisely this question for many years. Their chairman emeritus, Dr. Frank Drake, postulated an equation describing this probability, in 1961. It involves considering the total number of available stars (which, as Carl Sagan, the late astronomer and SETI board member, would have told us, is "billions and billions"), and then estimating how many planets will be orbiting them, the probability that each of those planets will be suitable for life, the probability that each such suitable planet will actually develop life, the probability that such life will eventually evolve some form of intelligence, and so on. When the number of available planets is multiplied by all those probabilities together, we get an idea of how many intelligent species might be waiting for us out there.

The problem is that we don't actually *know* any of these probabilities. What is the chance that a planet will be suitable for life? Or that life will actually evolve on a given suitable planet? Who can really say? Some have argued that, given the huge number of stars in the universe, surely intelligent life must have evolved somewhere else besides Earth. But what are the probabilities? The harsh reality is that, despite 40 years of intensive, systematic searching with advanced radio telescopes, no evidence of life elsewhere in the universe has emerged.

There is one slight exception. Recent analysis of fossil samples from our nearest neighboring planet, Mars, have indicated that maybe, just maybe, there was once microscopic life on that planet. It is even conceivable that all life on Earth is descended from early microscopic life on Mars, which "fell" to Earth in a meteorite. For philosophers, this possibility illustrates the fleeting nature of existence, and the possibility that we are all immigrants in ways even more profound than previously known. For science fiction and space buffs, it encourages further exploration of Mars and other planets and illustrates the possibility of truly alien life forms. But for probabilists, if life is known to have developed on not one but two different planets in our solar system, Earth may not be so unique after all. That increases the probability that life will develop on a given suitable planet, and also enlarges our sense of what planets are suitable. This, in turn, greatly increases the probability that other planets in the universe, besides Earth and Mars, can also support the creation of life. The probability of intelligent life elsewhere in the universe increases dramatically if it is true that there was once life on Mars. So, perhaps we are not alone in the universe after all.

The Sammy Sosa Incident

On June 3, 2003, Chicago Cubs baseball great Sammy Sosa broke his bat while batting in a run against the Tampa Bay Devil Rays. The game's umpire noticed a small bit of cork in a piece of the broken bat. The run that Sosa had batted in was immediately cancelled, and an investigation was launched into the "Sammy Sosa corked bat incident."

Corking a bat by drilling a small hole and inserting bits of cork can make it lighter and more springy, and thus potentially more effective. However, it is strictly against the rules of major league baseball. Sosa fully admitted to using the corked bat, but he claimed it was an honest mistake. He said the corked bat was intended only for batting practice, to entertain the fans with longer hits. In this one game, just this one time, he had mistakenly taken the wrong bat out during a real game. So the question became, was Sosa's corked bat a simple oversight or a deliberate attempt to cheat? Was this the only time Sosa had used a corked bat in a game, or was it something he did on a regular basis?

In Sosa's favor, he made no effort to hide or dispose of the corked bat in the seconds after it broke. Seventy-six other bats found in his locker later that night, plus several bats he had previously donated to the Baseball Hall of Fame, were all X-rayed by investigators and found to be perfectly legal, valid bats. On the other hand, Sosa had not been hitting well that season, which might have caused him to take desperate measures.

The debate raged on, driven by discussions of Sosa's seemingly honorable character, of other corked bats in baseball's history, of the need to please the fans, and so on. But from a Probability Perspective the question is, what is the probability that Sosa only used a corked bat this one time?

One way to approach this question is with p-values. The p-value is the probability that Sosa would have been caught, if he had used a corked bat only this one time. If this p-value is very small, it casts doubt on Sosa's claim that his bat corking was just a single, isolated incident. But if the p-value is not too small, Sosa's claim may be plausible. So, what is the p-value?

The key is that bats do not break often. Official statistics are hard to come by, but roughly speaking, an average major league baseball game has a total of about 75 at-bats (for both teams combined), and three or fewer broken bats. So, on any one at-bat, the probability of breaking a bat is probably no more than three chances in 75, or one chance in 25. By this reasoning, the probability of Sosa breaking a bat, and thus getting caught, on the *one* time he used cork, would be about one in 25, or 4%. This p-value is small enough to be statistically significant. By this reasoning, there is some statistical (though not completely conclusive) evidence against Sosa's having used cork just this one time.

Another issue that further reduces the p-value is that even once the bat breaks, it is not a certainty that the cork will be spotted. Indeed, the umpire who noticed the cork, Tom McClelland, happened to be the same umpire who had disallowed a home run by George Brett back in 1983 because of excessive pine tar on his bat (although that decision was later overturned). So, McClelland probably scrutinized bats and enforced the rules more strictly than many other umpires. With another umpire, Sosa might not have been caught. This suggests that the p-value, the probability that he would be caught if he used a corked bat just once, is even smaller than 4%.

There is one more point about p-values that might work in Sosa's favor. It is possible that corked bats, having been partially drilled and hollowed, are less solid than regular bats and therefore more likely to break. So it is possible that even if Sosa used a corked bat only once, he still had a high probability of being caught, because a corked bat has a high probability of breaking. If this is true, and the difference is significant, then the p-value might be significantly higher, thus supporting Sosa's contention.

It seems unlikely that corked bats are *much, much* more likely to break than regular bats; if they were, they wouldn't even be used during practice times. But from a Probability Perspective, Sosa's case all boils down to one question. If corked bats are, say, three times as likely as regular bats to break, the p-value could rise from about 4% to about 12%, a significant difference and one that would support Sosa's claim.

Sosa was eventually suspended for seven games, a substantial penalty but slightly lenient compared with other similar incidents in baseball history. However, the question of whether Sosa's actions were deliberate remains unresolved. All we can say for sure is that either Sosa was deliberately and repeatedly cheating *or* corked bats are much more likely to break *or* Sosa was extremely unlucky. It all comes down to probabilities.

Catching the Bad Guys

Low-probability events have another use: they are sometimes used to catch bad guys. We have all seen a movie or television detective whose suspicions are aroused because something "doesn't add up" or seems too surprising to be mere coincidence. Perhaps several seemingly unrelated bystanders all work for the same company, or the amount of money deposited in a criminal's bank account is precisely equal to the amount taken in a recent heist. When surprising or unexpected events occur, they can cause us to question or investigate further, hoping for the best while fearing the worst.

Modern companies use high-speed computers to identify exceptions to the norm in an effort to detect fraud and criminal activity. For example, telephone companies run computer programs to check if your long-distance phone charges fall within your "usual pattern." If suddenly you make calls totalling $1,000 to Sudan when you have never phoned Africa before, their suspicions are aroused. They may wonder if your phone line has been wired into by a black-market agent who is stealing and selling your long-distance services (apparently not uncommon). The phone company may inquire further, or in extreme cases, simply disconnect your line, in an effort to thwart the criminals.

Credit card companies also scan accounts for deviations from usual patterns. If you normally never charge more than $100 per month and then suddenly charge $8,000 worth of high-priced jewelry within a few hours, the company's computer programs will show an alert. The company might ask you to verify the purchases, or they may even invalidate your card until you can verify that the charges are legitimate.

Determining when charges should or should not be considered suspicious is a complicated question. If standards are too lax, many fraudulent charges may be logged before any action is taken. But if standards are too strict, much manpower may be wasted and many customers inconvenienced, all for nothing. The goal is to devise statistical algorithms that can distinguish between perfectly valid fluctuations in the spending patterns of honest customers and illegal activity.

Unfortunately, the algorithms don't always work perfectly. For example, I once needed to make two quick long-distance calls from a pay phone. I had just moved and my calling card number was not handy, so I charged the calls to a credit card instead. Late that evening, I returned home to a voice mail message from a bank, asking me to phone them immediately. I was told that there had been some suspicious activity on my credit card, which they feared might have been fraudulent.

Grimacing, I imagined spending days examining pages of charge sheets, trying to figure out which charges were really mine, and trying to convince the bank to remove the unwarranted charges. Resigned to my fate, I asked the bank representative for more details. There was a pause of a few moments while he accessed my files. Finally he came back on the line, slightly embarrassed. The computer had alerted him, he explained, because there were two unexpected phone charges of $5 each. That was the only anomaly, one that a human (unlike a computer) could see was hardly worth investigating. I guess their algorithm wasn't working so well that day.

Of course, probabilities can never prove that inappropriate activity was committed. When we arrive at a conclusion of criminality based on probabilities, the question we must ask ourselves is, what is the probability that we are mistaken? The justice system is sometimes too quick to minimize the possibility of mistakes because it uses faulty reasoning to estimate probabilities.

A Misuse of Multiplication

You're so angry—your flower bed has been trampled again. You assemble your four children and begin hurling accusations.

"One of you has trampled my flowers," you screech. "And I am going to figure out who!"

Looking from child to child, your eyes focus on Arthur, the oldest. "I think you did it," you accuse.

Arthur protests his innocence, but you silence him and continue your investigation. "Mrs. Chan from next door caught a glimpse of the act," you explain. "She didn't see who it was, but she was sure it was a boy.

Furthermore," you add for dramatic emphasis, still glaring at Arthur, "you are a boy. Since only two of my four children are boys, the probability is 50% that the perpetrator just happened to be a boy too, purely by chance, without you being guilty of anything."

Arthur feels aggrieved, but figures that 50% is still a pretty large probability. Maybe he will get off.

"There's more," you continue. "Mrs. Chan also said the child had blond hair. Only three of my four children have blond hair, so the probability is 75% that the perpetrator just happened to have blond hair too, purely by chance."

Arthur is getting nervous, but you continue. "Mrs. Chan also saw that the flower stomper was wearing a blue jacket. Only two of my children have a blue jacket. That's another 50%."

Arthur is fidgeting, but doesn't dare to run. "And, finally," you declare, "the perpetrator had to be able to climb into the garden. Lisa and Jennifer are both too small to do that. Only 50% of my children could have pulled off this maneuver."

Arthur starts to reply, but you cut him off. "Don't interrupt! I'm conducting an investigation here!" You pick up a handy calculator and begin pressing buttons. "Let's see, 50% times 75% times 50% times 50% equals . . ."

The air is thick with silence as you do the calculation. Finally you conclude, "Just 9%; there is only a 9% chance that you just happened to fit the profile of the perpetrator, purely by chance. That's a small enough probability for me. Go to your room, and don't come out for five months!"

Hurt and defeated, Arthur trudges upstairs to his room. Meanwhile, just out of view, your other son, Jonathan, is smiling. He looks resplendent in his blond hair and blue jacket. He runs outside to enjoy his freedom.

In fact, it was unfair to multiply all of these probabilities together. The reality is that both Arthur and Jonathan were agile boys with blond hair and blue jackets. So, the probability that Arthur fit the description by pure chance was 50%, not 9%. Without a doubt, Arthur was unfairly caught in the uncompromising web of over-eager parental justice.

Similar problems arise when DNA profiling (also called "DNA fingerprinting") is used in criminal proceedings, such as during the infamous O.J. Simpson trial. DNA is our personal genetic code, and except for identical twins, every person's DNA is unique. However, current DNA profiling technology does not match the entire DNA sequence. Rather, only a small number of DNA "markers" are identified and matched. If samples from the suspect and from the crime scene have the same markers, this provides evidence of guilt. But how much evidence?

The probabilities associated with DNA profiling were controversial and frequently debated, especially during the early years of using DNA in criminal trials (the late 1980s to the mid 1990s). One issue was whether it was valid to multiply the probabilities corresponding to the different markers. Are the various markers "independent," so their probabilities can be multiplied, or are they "dependent," so that such multiplication is invalid? Prominent statisticians supported each side of that debate while criminal convictions hung in the balance.

A related issue is that DNA profiling attempts to compute the probability that a randomly chosen person would happen to match a specific DNA sample, just by chance. But chosen randomly from what population? From the world at large? From people who live near the crime scene? From people of the same race as the suspect? The choice of which comparison population to use can significantly affect the probability of a match.

Even if a suspect's DNA matches, and even if there is a very small probability that a randomly chosen person's DNA would match, does that mean the suspect is guilty? Maybe he is innocent regardless. After all, out of all the world's population, it is not surprising that *someone's* DNA would match the suspect's by pure chance. DNA profiling attempts to compute the probability that a randomly chosen person's DNA would match that taken from the crime scene. But what is really important is the probability that the suspect is guilty, which is a different matter and harder to quantify.

The O.J. Simpson murder trial provided an extremely high-profile battlefield for this question. Blood samples matching Simpson's DNA had been found near the two victims' bodies, and blood samples matching the victims' DNA had been found in Simpson's car and on a glove behind his home, evidence that seemed to point strongly to Simpson's guilt. Both the prosecution and the defence legal teams called numerous statisticians to debate such technical matters as *likelihood ratios* and *frequencies* and *mixtures,* all in an effort to determine the probability that these blood-sample matchings had occurred by pure chance. One prosecution witness, Robin Cotton, concluded that the probability was less than one chance in 170 million that a randomly chosen person's DNA would have matched the DNA of a blood splotch found near the victims' bodies, as Simpson's did.

Some extra statistical courtroom drama was provided when one of the prosecution's statisticians, Dr. Bruce Weir, was forced to admit he had made a mistake during some last-minute court-mandated extra calculations. Dr. Weir quickly corrected his calculation, and the probability of a match by pure chance remained extremely small. However, the miscalculation may have lessened the credibility of the DNA evidence, and Weir admitted that "I'm going to have to live with that mistake for a long time."

In the end, the DNA evidence against Simpson wasn't so much discredited as discounted. One of the police investigators who had handled some of the evidence, Detective Mark Fuhrman, had previously been recorded using racial epithets. The defence team argued that Fuhrman was biased against African-Americans, and this led them to suggest that the police may have planted evidence, which would make all of those one-in-170-million probabilities completely irrelevant. It seems likely that this suspicion, more than any one other factor, led to the jury's verdict of not guilty. One of the jury members said in a post-trial interview, "I didn't understand the DNA stuff at all. To me, it was just a waste of time. It was way out there and carried absolutely no weight with me at all." It seems possible that this particular jury member was lacking in Probability Perspective.

The controversies around DNA profiling continue to this day and remain an active area of statistical research. Nevertheless, most statisticians would agree that if DNA taken from a crime scene is found to match that of a suspect, it provides fairly strong evidence that the sample came from the suspect, assuming the sample was analyzed properly and was not planted or tampered with.

For another perspective on using probability to catch bad guys, we turn to the adventures of Ace Spade, PPI.

9

Interlude

The Case of the Collapsing Casino

As a half-time interlude, let's follow the fun adventures of Ace Spade, PPI. (Warning: Serious, sober-minded readers may wish to skip this chapter.)

* * * * *

It was a cold winter day. As cold as the logic of a mathematical proof. The window rattled in the wind, in time to the clicking of Doris's typewriter. "I'll be in my office, Doris," I snapped. "Okay, Ace," Doris chirped. "I'm almost finished logging the office expenses!"

Doris was chirpy, all right. Too chirpy. She knew as well as I did that business hadn't been good. Nobody bothers with the Probability Perspective anymore. Doris was dependable, but adding up the books would take less time than it takes an electron to return to its low-energy state.

From inside my office I heard the phone ring. I held my breath; could this be the client I needed? "Office of Ace Spade, Probabilistic Private Investigator," I heard Doris sing out. "How may I help you?" After a pause, I heard Doris say, "Please hold, I'll see if he can squeeze you in." A meeting was being arranged!

I could squeeze the caller in, all right. Doris knew I had no other appointments that day. She soon appeared in my office, eyes gleaming behind her rectangular black glasses, to confirm a meeting with Jenny Jupiter from the Baker Betting Building.

Late that afternoon, I heard someone arriving. Seconds later, Doris knocked on my office door. "Ace, this is Jenny Jupiter." In walked Jenny, and it was everything I could do to keep from falling over. Jenny was a looker, all right. Bright-blond hair framing deep blue eyes and pouting lips. Legs as long as a boring calculus class. A sweater stretched out in the shape of a giant infinity sign. I steadied myself against my desk and tried to keep my composure.

"Er, uh, sit down," I stuttered. Jenny quickly sat, but her expression showed no pleasure. "Oh, Mr. Spade," she began, "you've got to help me! It's my husband's—I mean my fiancé's—betting parlor. It's all gone so wrong! Just when things were looking up . . . just when George was getting back on his feet . . . just when it looked like our wedding might finally go ahead. . . ."

I suppose I should have comforted her, but I didn't trust myself to get within a 1.5-meter radius. Instead, I tried to keep things professional. "My fee is $100 an hour," I began. It's always best to settle the fee while the client feels distraught. When she nodded, I continued, "Now tell me the details."

"Oh, Mr. Spade," she said again, "the Baker Betting Building—that's my husband's place—was doing just swell. The bank balance was climbing, the customers were having fun. George and I went ahead and scheduled our wedding for next month. We figured we could afford a big party, everything included. But now. . . ." She dabbed her eyes.

"Go on," I said as neutrally as I could. She seemed upset, but her manner was nevertheless focused and earnest, as if she knew just why she was here. I can respect that; know your equation before you solve it, I always say.

She composed herself, resting her hands on my desk, showing a big, flashy diamond ring. "In the last few months, things started turning around," she continued. "Customers started winning their bets. The bank balance went down. And then in the last few days, two high rollers won big, really big." She paused for effect, and looked me straight in the eye before continuing, "It's just about cleaned us out, Mr. Spade!"

I forced my glance away from her sweater and tried to think. It didn't add up, see. With casinos, the fix is in. The odds are set. The scales are tilted in the house's favor. Sure, a few high betters might score now and then. But in the long run, the house wins, see. That's the way it is. It's the Law of Large Numbers. It's in the bag.

Jenny stood up, and reached out her hand to thank me. I found myself saying, "Why don't I follow you back now?"

We stepped out into the freezing wind. My car was in the shop, as usual, so she drove. Her back seat was littered with financial documents, a makeup kit, half a tuna sandwich wrapped in clear plastic, a cheap paperback romance novel, and a Greek Islands tour book with an envelope sticking out. Ah, Greece, the birthplace of geometry, the home of Euclid and Pythagoras and Archimedes and—

Jenny must have seen me staring. Shivering, she said, "You know, in Athens yesterday, it was a nice, warm 21 degrees Celsius!" Glancing at her foot on the accelerator, I realized I could use a little warmth myself.

It had snowed a lot the previous day, so driving was slow, but finally we reached the Baker Betting Building. Jenny walked up the steps, and I followed closely behind, more distracted than a mathematician finishing off a proof during a cocktail party.

"This is the place, Mr. Spade," she indicated, waving her hand around. I took a good look. It was like a thousand other casinos—bar at the back, tinny jazz music coming from a lone speaker, smoke as thick as a trigonometric identity. Customers milled about, playing blackjack, poker, slot machines, and roulette. Some looked pretty flashy, others as down on their luck as a lottery addict.

Jenny took me to a back office. "This is my husband, George Baker," she said, introducing me to a large, square-shouldered fellow with a crooked nose. "I mean my fiancé," she corrected herself with a coy smile at Baker, who smiled back.

"Beautiful, ain't she?" Baker asked as Jenny left.

"I hadn't noticed," I replied coldly.

"We're gonna get married next month," he went on. He seemed as lovestruck as a graduate student who's just chosen his Ph.D. thesis topic.

"Down to business," I snapped. "Jenny says you've been losing money lately. I need the details. All of 'em."

"It's true," Baker sighed. "Why, just last week, two different players won the three-cherry jackpot on the slot machine—that's a $20,000 payout! I can't afford two of those in a row! It's ruining me! Usually, there are only about eight three-cherries all year. Something is wrong!"

It sounded bad, all right. "Those two players with the three-cherries," I asked, "are they here now?" They were. Baker pointed them out with more than a little anger in his voice. One, Johnson, still playing the slot machines, wore a suit as shiny as a two-headed coin. The other, Alberts, had moved over to the poker table, where he was losing badly to a shifty-eyed youth named Richards.

"Ah, Shifty Richie," Baker grumbled. "When you think his hand is strong, he's bluffing. When you think he's bluffing, he's got the juice. The kid never loses." Something about what he said stuck in my craw. Never loses, huh? Nobody never loses.

I looked around some more. At the roulette table, a nervous-looking young man was spinning the wheel and collecting the bets. As the wheel wound down, he would squeal, "No more bets, please!" Baker explained that normally the roulette wheel was Jenny's job, but Frankie was working today because Jenny was off, first visiting me and now balancing the books in the back room. Around the roulette wheel were a few lonely looking losers, including a fidgety middle-aged sad sack with a loosened tie and a dirty shirt. "Name's Alpha Beta," Baker explained. "Comes in nearly every day and bets just $10 each spin—always on black. Guy's so insecure that half the time he backs out and doesn't bet after all," he continued with a chuckle.

At the blackjack table, the dealer was a pretty brunette named Lisa. She dealt like a real pro, crisp and clean and as fast as a Beowulf computer cluster. Baker said she'd been working there about two years, did a good job, and earned her large bonus payments, but he didn't know much about her past. At the moment, she was cleaning out a tipsy business fellow who bet high and split whenever possible but never seemed to win. Baker said his name was MacDonald, and he came in twice a week, always played blackjack with Lisa and always left looking sadder than when he arrived.

A thought occurred to me. "You keep your daily totals for the different slot machines and tables?" I asked. "Sure!" Baker snorted. "What kind of a fleabag betting parlor you think I'm running here? I never read that stuff myself but, sure, we got it." We went back to his office, and he called to Jenny to bring in the books she was working on.

Once we were seated, he asked me, "What is this all about? Do you have things figured out?" Actually I did have the beginning of an idea. But just then Jenny arrived, carrying a tray containing the financial books plus two steaming cups of coffee. She put the books down on Baker's desk, then leaning over she carefully placed one cup in front of Baker and the other in front of me. Her work with the financial figures hadn't changed her own, and I lost my train of thought. "Er, uh, thanks for the coffee," I offered feebly.

Jenny gave a smile that could warm up absolute zero, and then walked away. Both Baker and I stared after her. Only once she left the room did I recover and turn to the books.

"So, what's your big idea?" Baker asked, taking a sip of his coffee. "I mean, with what we're paying you, we need some answers!"

I looked over the books. There were totals, day by day, for the amount won or lost at each slot machine and at each betting table. Sure enough, it confirmed what I suspected. Things were starting to fall into place. "I think I've got something," I told Baker.

Excited, Baker stood up. "I just knew you would! Now we'll get those bastards!" He stepped out of the office and called, "Johnson! Alberts! Get in here!" Jenny was just walking past the now empty roulette table, and Baker said, "Jenny, perhaps you'd better hear this, too."

As Baker returned to the office, he stumbled a bit and slumped quickly back into his chair. "Whoa, I don't feel so great," he mused. "I'm kinda dizzy. It's funny, I felt fine just a minute ago."

The others quickly arrived. "What's this all about?" demanded Johnson in a huffy voice, his suit as shiny as ever. "I'm up over a grand at that slot machine, and I'm not stopping now! Customers have certain rights, you know!" Alberts was more sheepish, muttering, "Perhaps a break from Shifty Richie will do me some good." A few other customers, including MacDonald and even Richards, stuck their heads in the door to see what was going on. But my eyes were on Jenny, sauntering slowly into the office and leaning against a wall by the door.

Baker's face looked pale, but he managed to mutter, "Spade, tell them about . . . about the . . ." before trailing off. Feeble as his direction was, I took it as my cue.

"Listen up," I barked. "This joint has been losing money hand over fist, and it's time to find out why." This was met with choruses of "That's not my problem" and "Just what are you implying, mister?" but I ignored them and went on.

"In the past week, two customers," I looked squarely at Johnson and Alberts, "have each won a three-cherry $20,000 payout. That should only happen about eight times a year. What are the odds of two wins in one week?"

"Damned if I know!" growled Johnson. "Frankly, damned if I care, too!" He started to walk out the door.

"Not so fast," I shot back. "I'll tell you the odds." My listeners sat back and sighed, resigned to the lecture. "Eight times a year corresponds to an average of 0.154 wins per week." This elicited a few groans about "how can you have 0.154 of a win?" but I continued. "Since the different wins are independent of each other, that means the probability of two wins in any one week is," I could feel everyone lean in for the answer "just over one percent." Everyone was frozen now, wondering how to interpret this figure. "Oh, that's unlikely," I conceded, "but not so unlikely that it will never happen. In fact, it should happen about one week every two years—that's Poisson clumping for you—so it's about time that it did."

There was a moment's silence while everyone took in what I had said. Somebody asked if poisson was French for "fish." But gradually it dawned on them all that I was letting the slot players off. This caused them a brief glimmer of pleasure, but then their annoyance returned. "You mean you called us in here just to tell us that?" demanded Johnson.

Even Baker, woozy as he was, still managed enough anger to wail, "With what we're paying you, that's the best you can come up with?"

I tried to regain control. "The three-cherry wins are what threw us all off," I explained. "I was puzzled, since your slot machines are serviced regularly and can't be tampered with. So in the long run, they should be in favor of the house, as they're designed to be. But your downfall didn't come from the slots. It didn't fit." Baker seemed puzzled, so I continued. "Once I realized that two three-cherries in one week wasn't that unlikely after all, I decided to look elsewhere."

I had my listeners entranced, so I went on. "Next I wondered about the poker table. How could Richards always win? I figured probably he cheats." At this, Richards jumped up, raised his fists, and cried, "How dare you!"

"Relax," I told him. "It doesn't matter. In poker, the house takes a cut off the top of every hand, so no matter who wins, the house always comes out ahead." Richards still looked ugly, so I added, "Seeing as the house is my client today, I don't care if Richards pulverizes his competition like a proof by contradiction." They were puzzled by this remark—nobody understands logical proofs these days—but even so, Richards seemed mollified and dropped his hands to his sides. As he did, I saw the corner of a King of Diamonds stick out from his left sleeve. But the poker players weren't the ones paying for my services, so I said nothing.

"That brings us to the blackjack table," I announced. "That Lisa is the smoothest dealer I've ever seen. Bet she was a hustler before working here. Deals so fast I bet she could lay any card down anytime, and no one would be the wiser." Lisa, who had stuck her head in the doorway, scowled, unsure whether to be offended by my implication or flattered by my assessment of her abilities.

"Her smoothness didn't stop at cards, either. At first I was puzzled about why poor MacDonald kept on betting every day, even though he was losing. Then I saw it. It was the one thing that probability can't predict. Love! He had fallen in love with Lisa!" At this MacDonald looked mortified, his head buried in his shoes like an electron in a potential well. Mixed in with Lisa's discomfort, I saw a hint of a smile. The other players looked at her accusingly. Bizarrely, an image of my secretary Doris flashed through my mind.

"Again," I told them, "it doesn't matter. Whatever Lisa has done in the past, now she's working for my client and making lots of money for him, too. She's earned her bonuses, friends." The players stared at Lisa with a newfound horror, and they variously vowed never to play blackjack again. Baker, still slumped in his chair, seemed both impressed with Lisa's operation and worried that my revelations were driving away his customers.

"This brings us, finally, to the roulette table. That's right, the poor, old, underused, ignored, disparaged roulette table." Several players scoffed at my mention of roulette, and even Baker seemed disappointed. But it was Lisa, still full of contempt, who spoke up. "Oh, come on now," she spouted. "We get that roulette table tested and serviced just the same as the slot machines!" The players murmured agreement.

"That may well be," I conceded, "but there's more than one way to compute a double integral." Again they looked puzzled, so I went on. "At first I dismissed roulette. The players never bet more than $10 at a time, usually just on red or black so the payouts are just $10 at a time. Furthermore, the odds are all fixed in favor of the house, so in the long run no player can ever make money off roulette. No way. Impossible." Everyone agreed. "Exactly!" exclaimed one.

"That is," I spoke up above the din, "no one can make money by playing fairly." This brought fresh howls of protest. "I already told ya, the wheel gets tested!" Lisa practically screamed.

"Oh, I'm not thinking about the wheel. I'm thinking about the bets." I had their attention now, so I went on. "It all came to me when I heard Frankie cry out, 'No more bets please!'" This got a bit of a laugh, as Frankie's repetition of this line had gotten on everyone's nerves a little. "I thought, what if there were no cut-off on bets? What if you could change your bets after the ball had landed?"

At this Frankie, who had been shyly listening from the other room, pushed his way in, his face red. "You listen to me, mister!" he wailed. "I know some people don't like the way I give instructions. But I do my job proper, and I would never, ever let a fella change his bet once the ball had landed!"

"Hey, hey, Frankie," I said gently. "Nobody's accusing you of anything. I know you did your job properly." This stopped Frankie in his tracks, and I swear he started to sob quietly from the built-up emotion. Fortunately for all of us, he left the room.

"No, not Frankie," I told the others. "Frankie's a good kid. But he was just filling in. You all know who's the usual roulette spinner." They certainly did. At my words, everyone turned to stare at Jenny, at least those who weren't eyeing her already.

Jenny played it cool. Too cool. "Why, Mr. Spade," she said with a smile, "I sure hope I haven't done anything wrong. If there's anything I can do to help catch whoever's been doing my husband—I mean my fiancé—in, then you just give the word, fella!" She looked at me with those deep blue eyes, her hair falling gently over her cheeks, and I had a moment of hesitation. But I forced myself to look away and go on.

"Jenny couldn't mess with the wheel, that would be too obvious. What she could do is to allow a confederate to withdraw his bet after the ball landed now and then, if he saw he was going to lose. That way, in the long run, he would have fewer losses than wins and come out ahead."

This elicited more howls of protest. Even Baker, who was now listening with rapt attention, wondered how a few $10 bets could make much difference either way. "Oh, sure, it wouldn't make much difference at first. But let's say this confederate came in every day, stayed for eight hours, and bet on black, once every 30 seconds. Then 18 times out of 38 it would come up black and he would win $10. As for the rest of the time, suppose this confederate was watching for the image of a dark ball on a red spot. Suppose that even *half* of the times that it came up red, he managed to withdraw his bet. Then on each roll, instead of averaging a loss of 53 cents, he would average a win of—" I computed fast "—of $1.84. After a month, he would be up—let's see—about 54 grand."

This last figure really threw them. How could a few puny $10 bets add up to $54,000 a month? Repetition and quantity, that's how. Small changes in the short run, repeated over and over, add up to big changes over time. Big enough to bring down Baker. One player had the gall to question my calculations, but I shot back, "I'm a Probabilistic Private Investigator, pal. I'm a professional."

"By the time I checked Baker's books," I continued my narrative, "I knew what I was looking for. And sure enough, the roulette figures are in the negative almost every day for the past four months. And that, my friends, is as rare as the steak down at Chez Jacques." They were getting more interested now. "Oh, it's just a few thousand here and a few thousand there, but it adds up. And the thing is, since Jenny kept Baker's books, no one else noticed these odd figures. In fact, Baker himself didn't know there was trouble until the unexpected three-cherries run, when suddenly he couldn't pay his bills." People were now looking at Jenny with a mixture of anger and pity. I knew they'd find it hard to believe she was guilty. Unfortunately, I wasn't done yet.

"The hardest part was identifying Jenny's confederate. Then it hit me. She had a Greek Islands tour book in her car, and she knew the temperature in Athens. She was planning a one-way trip to Greece. The envelope in her tour book could only be one thing: an airplane ticket. And with whom would she be travelling? Why, probably a Greek. Someone like Alpha Beta, who must be Greek because his name is the first two letters of the Greek alphabet!"

At this Baker tried to get to his feet. "Beta! Beta!" he cried out. "Get in here!"

"Oh, Beta's already left," I explained. "He cleared out from the roulette table as soon as Jenny gave the signal. Jenny herself was on her way out when you called her back. They had their escape all planned, and they were taking Beta's roulette winnings, plus Jenny's big diamond engagement ring, with them. All Jenny's talk about 'my husband, I mean my fiancé' was designed to throw us off. She was as likely to marry Baker as a coin was to come up heads 100 times in a row."

My audience puzzled that one over, and Baker himself looked heartbroken, but I quickly went on. "When the three-cherries unexpectedly hit and caused a premature financial crisis, they had to make a quick run. They probably planned to leave yesterday, but the airports were closed, due to the snowstorm. So instead, Jenny followed Baker's instructions and called me in. She was just stalling for time, figuring she could escape before I'd learn anything." I grinned and looked around. "She got that part wrong."

The players couldn't believe that someone as beautiful as Jenny would run off with someone as strange as Beta. It was time to move on to my last pronouncement. The home-run hit. The QED. My Ace in the hole. My Ace of Spades, so to speak.

"There was something else," I explained. "Just when Baker called you all in here, he started to feel weak and dizzy." A quick glance over at Baker confirmed that he was still in this state. "Yet he'd been alert and energetic just before. What had changed? Then I figured it out: Jenny had served him coffee, and he'd had his first sip just before going weak. Coincidence? Not likely. Out of all the 1,440 minutes in a day, why would he go weak the very minute he drank Jenny's coffee? That stacked the odds. That was the clincher. Jenny had drugged Baker, making him confused so he wouldn't notice her escape. I bet she was trying to drug me, too—that's why I didn't drink anything." Looking down at the cups on the table, I continued, "I'm sure the lab boys will find that these coffees have been spiked."

Actually, I didn't know any lab boys—I'm a probabilist, not a chemist—but I figured somebody somewhere would know what to test for. Jenny must have figured this, too, because she suddenly reached out to knock the cups over. I saw her in time and grabbed her wrists first.

So there I was, face to face with the prettiest dame I'd ever seen. She was crying gently, which only made her eyes look more blue, her lashes longer. I felt more rotten than a discarded ham sandwich in the math student lounge. Sometimes I wondered why I'd ever become a PPI, instead of getting a respectable professorship at a nice university some place.

The police rounded up Jenny and Beta without trouble, and they were sentenced to more years than it takes to do a doctoral dissertation. Baker's business thrived again, and he was so grateful that he paid me double my fee.

As for me, I discovered that it's not just in mathematics that the simplest ideas are the best. Sometimes what is nearest is also dearest. Doris and I fell in love, and we were married in a small, quiet ceremony. Our PPI business is still hobbling along. Baker attended our wedding, and we see him now and then. Sometimes, when he's in a good mood, he invites us over to his casino to play a little roulette. Every once in a while, with a wink, he lets me withdraw a losing bet on black, just to show Doris how it was done.

* * * * *

So ends the tale of Ace Spade, PPI. Beauty, riches, poison, deception, resolution—probability theory has them all.

10

Fifty-one Percent to Forty-nine Percent

The True Meaning of Polls

A number of national elections were held in 2004, including in Spain on March 14, in Canada on June 28, in Australia on October 9, and in the United States on November 2. The run-up to each of these elections featured large numbers of polls that attempted to gauge the opinions of the citizenry and predict the upcoming results. Most of those polls asserted various "margins of error" for their conclusions. For example:

- One month before the Spanish election, the Center for Sociological Research completed a huge survey of 24,000 Spanish voters. They predicted that the governing Popular Party would receive 42.2% of the vote, while the opposition Socialist Party would receive 35.5%. They asserted that their results had a margin of error of 0.64%.
- Two days before the Canadian election, EKOS Research Associates sampled 5,254 Canadian voters and predicted 32.6% support for the Liberals, 31.8% for the Conservatives, and 19.0% for the New Democratic Party (NDP), results they claimed "are considered accurate to within 1.4 percentage points, 19 times in 20."
- Two weeks before the Australian election, an ACNielsen poll surveyed 1,397 voters and showed the governing coalition of Prime Minister John Howard leading the opposition Labor Party by 52%

to 48%, with "a margin of error of plus or minus 2.6 percentage points."
- Eleven days before the United States election, Reuters/Zogby surveyed 1,212 likely voters, giving George W. Bush a lead of 47% to 45% over John Kerry; they said that their margin of error was plus or minus 2.9 percentage points.

Similar claims are made all the time. What do they mean? Can polls really predict who will win an election? How closely do their results reflect the true opinions of the electorate? On what basis do they assert their accuracy or error levels? Are they really sure?

And why does it matter what the polls say, anyway?

Public opinion polls have a special place in the modern political process. Decisions are made, policies are introduced, performances are evaluated, and election campaigns are run, all based on polling results. Politicians may claim that they "don't believe in polls" or that "the only poll that counts is on election day," but the reality is that they make many of their decisions according to the latest (and perhaps secret) poll numbers.

It could be argued that polls provide us with more direct democracy than elections do. In elections, we get to vote only for a party or candidate—and often all parties behave similarly on certain issues. But with polls, our opinion about each particular issue is taken into account, albeit indirectly, as politicians plot their next move.

At times, polls seem to provide an extra level of communication among the electorate. For example, Toronto's 2003 mayoral election featured 44 different candidates, at least five of whom were well known and high profile. The candidates held a wide variety of political opinions, thus creating a jumble of confusing voting possibilities. However, over the course of a long and bruising campaign, during which polls were published regularly, two candidates emerged from the crowd: a respected center-left candidate (who won with 43% of the vote), and a respected center-right candidate (who came a close second with 38% of the vote). The other candidates (including the original leader in the polls) each took less than 10% of the vote. Citizens had used the polls to create a dialogue of sorts, and to narrow down a huge field of mayoral candidates into a clear, compelling choice.

In the British election of 2005, Prime Minister Tony Blair spoke for most analysts when he concluded that "it is clear that the British people wanted the return of a Labour government but with a reduced majority." However, the choice "reduced Labour majority" did not appear on any election ballot. Rather, the British voters used poll results to balance out their votes and achieve their desired election result.

Another example of the same phenomenon was the Quebec sovereignty referendum of 1995. The province of Quebec was voting on whether or not to secede from Canada and form its own sovereign country. The stakes were high, and many Quebecers were in a quandary. A clear Yes vote would lead to quick independence, which most Quebecers did not want. But an overwhelming No vote might weaken Quebec's bargaining position, and cause the rest of Canada to discount and ignore Quebec's constitutional and financial demands over the following months. Each individual Quebecer could only vote a simple Yes or No. But by following the polls carefully, and switching their allegiance as needed, Quebecers managed to achieve the balanced vote—49.4% Yes and 50.6% No—that many of them sought. Without polls, such a close result probably wouldn't have been achieved.

An on-line survey by *Maclean's* magazine in June 2004 found that fully 91% of respondents answered No when asked if poll results affect their vote. But I don't believe them. I think that many voters are indeed influenced by polls, though they may not realize or admit this fact.

The Meaning of Polls

It seems clear that polls have a significant impact on our society and political system. But, what exactly do they mean?

Suppose a polling firm samples 1,000 people and asserts that their results are "accurate within 3.1 percentage points, 19 times out of 20." It may seem at first that they are claiming that, 19 times out of 20, the results of the next election will be within 3.1% of the firm's numbers.

In fact, making such an incredible claim would require a precise understanding of how political opinions change between the poll date and the election date; the extent to which citizens will say one thing to pollsters and then vote differently; the probable actions of undecided voters and of those who refused to respond to the survey; which citizens will or will not bother to vote; and a host of other intangible factors.

Pundits and analysts work overtime trying to understand these factors, and statistical modeling is used to try to estimate them. However, voting intentions are very complicated and subtle, and polling firms do not make claims of great predictive powers in these areas.

No, what the pollsters are claiming is something far more mundane. They are claiming that, if they conducted a "comprehensive poll" by phoning every single eligible voter in the province (rather than just 1,000), then there are 19 chances in 20 that the results of their poll of 1,000 people would be within 3.1% of the results of the comprehensive poll.

Or, to put it differently: If they did 20 similar polls in a row (i.e., each time again phoning 1,000 adult residents at random and asking them about their voting preferences), about 19 of them would be within 3.1% of the "right" answer (i.e., of the true political preferences that residents will report to pollsters at that particular time).

In short, polls don't have any magical powers. They don't necessarily predict what people will actually do, or how they will actually vote. They can only report on how people answer questions on the telephone. And the poll's margin of error isn't directly concerned with who will actually win the election; it merely describes how accurately the poll is measuring what all of the eligible voters would say if they were all asked the same question.

Another difficulty with polls is that the political situation can change, sometimes very quickly and dramatically. Pollsters try their best to anticipate such turns. For example, they attempt to ascertain the "depth" of a candidate's support by asking such supplemental questions as, "Would you continue to support your candidate, even if he performed poorly in the upcoming debate?" or, "even if on some issue, he took a position that you did not agree with?" But such efforts are of limited value and cannot always predict the outcome of an election. In any case, such considerations are not reflected in any poll's "margin of error."

The most spectacular example of changing voter opinion was the 1948 election for President of the United States. Early pre-election polls had predicted an easy win for Republican Thomas Dewey over Democrat Harry Truman, by an expected margin of victory of between 5% and 15%. The result seemed so clear that polling companies didn't bother to do any late polls to check for last-minute shifts in voter preferences. The result was a narrow victory for Truman, huge embarrassment for the polling companies (about 30 American newspapers immediately cancelled their subscriptions to the Gallup polling results), and so much confusion that the *Chicago Daily Tribune* published a large headline "Dewey Defeats Truman" just as Truman's victory was established.

The polling companies have learned their lesson since 1948, and now routinely conduct polls until just a day or two before the election. But even this does not completely eliminate surprises. In the 1992 United Kingdom general election, opinion polls gave Neil Kinnock's Labour Party a slim lead over John Major's Conservative Party. However, when the votes were finally counted, the Conservatives had eked out a majority (of just 21 ridings, out of 651 total). The Conservatives then continued to govern until 1997. My British statistician friends were doubly disappointed by the 1992 election results: they were disappointed that the Conservatives (who were seen to be unsupportive of universities) were re-elected, plus they were embarrassed that the opinion polls had gotten it wrong. Some commentators blamed an election-day anti-Labour headline in the populist tabloid the *Sun* for the crucial last-minute opinion swing.

In the days before the 2004 Canadian election, that EKOS poll and several others all showed the race to be extremely close, with the Liberal and Conservative parties each at around 31% or 32%. And yet, on election day, the Liberals received 36.7% and the Conservatives 29.6%. The

Liberals went from being virtually tied with the Conservatives to handily beating them. What happened? It seems that in the final day before voting, about 5% of voters, fearful of a Conservative election victory, shifted their vote to the Liberals, largely from the NDP. ("The Liberal surge was literally overnight," lamented a senior NDP adviser.) This shift was enough to give the Liberals a solid victory instead of a tie. In the aftermath, many people blamed the polling companies for incompetence, but in reality they were simply the victims of last-minute, unpredicted voter shifts—in other words, of democracy.

The Spanish election of 2004 was more controversial. Three days before the election, terrorists bombed four commuter trains, killing 191 people. Government officials immediately blamed the Basque separatist group Euskadi Ta Askatasuna (ETA), although al Qaeda was later found to be responsible. The government ended up being blamed for joining the American invasion of Iraq the previous year and also for misleading the public over the train bombings. The result was a large voter shift away from the government, from a comfortable lead in pre-election polls to a solid defeat on election day. And once again, no poll's margin of error had predicted this dramatic turn of events.

Another difficulty for pollsters is that people don't always tell the truth. Some citizens are simply dishonest about their voting intentions. If dishonesty occurs in a random fashion, without any particular pattern, the poll's results remain valid. However, if the respondents' dishonesty is weighted in a particular direction, pollsters are in trouble.

One example is votes for African-American electoral candidates in the United States. Many people do not wish to appear racist, and they will tell pollsters (and others) that they intend to vote for an African-American candidate. However, on election day, some of them will switch to another candidate instead. The net result is that polls tend to slightly overestimate the support of African-American candidates. Just ask David Dinkins. In 1989 he was predicted by polls to have an easy victory as mayor of New York City, but in fact he won very narrowly. Then in 1993 he was predicted to narrowly win re-election, but in fact he lost narrowly to Rudy Giuliani. Similarly, in 1996, African-American candidate (and former

Charlotte, NC, mayor) Harvey Gantt was predicted to edge out Senator Jesse Helms, but in fact he lost by several percentage points.

In every election some voters say they are "undecided." Usually undecided voters are simply not counted in the poll's results, and for the most part this doesn't cause any problems. However, if for some reason undecided voters tend to behave in a particular way, they can skew the results. An important example is, again, the issue of Quebec sovereignty. It is now known that, when discussing Quebec sovereignty, many Quebecers declare themselves to be "undecided" because they plan to vote No in a referendum, but they are getting pressure from their friends and neighbors to vote Yes. (As Jean Chrétien remarked about the 1980 referendum, "It was easier for the federalists [No voters] to keep quiet until they got into the voting booth.") When polling companies measure support for Quebec independence, they usually allocate 75% of the undecided vote to the No side, and just 25% to the Yes side. If they didn't make this correction, their polls would overestimate support for sovereignty.

Some voters will simply refuse to talk to pollsters. Perhaps they are private people by nature. Or maybe they are too busy. Perhaps they object to the very idea of pollsters. Or they are never home to answer the phone. Or, in the newest twist, they use only a cell phone (which pollsters are forbidden to call because the cell phone owner may have to pay for incoming calls), and are thus unreachable in telephone polls. In any case, if the unreachable voters have the full spectrum of opinions, without any particular bias toward certain positions or candidates, inability to poll them will make little difference to a poll's accuracy. However, if unreachable people tend to vote a certain way (for example, perhaps recent immigrants are less likely to confide in pollsters, but are also more likely to support a particular party), then the poll results can be skewed. Some pollsters fear that, as more and more people become fed up with unsolicited phone calls, or switch to using only a cell phone, this problem will increase.

Many people who claim to support a particular party will not bother to vote on election day. Thus, a big part of modern political organizing is arranging volunteers to remind supporters that it is election day, escort supporters to the polling stations, and otherwise "get out the vote." Once again, if all parties are equally successful at getting out the vote, a poll's accuracy will not be affected. But if one party is significantly more successful than another, they could end up with a higher percentage of the actual votes cast than was predicted by pollsters.

The Sad Tale of the Apathy Party

After careful consideration, you decide that the best possible lifestyle is one of apathy. You dream of a nation of couch potatoes, each constantly slumped over watching low-quality television programs. To make your dream a reality, you inaugurate the Apathy Party, dedicated to encouraging indifference throughout the land.

Pre-election polls indicate that you have struck a chord. Fully 40% of citizens tell pollsters that they support your policies. An election victory seems within reach.

On election day, you are shocked to find that not a single citizen cast a vote for your party. It seems that all of your supporters stayed home, watching television.

Pollsters have to be especially careful when conducting surveys about controversial questions or illegal activities, like drug use. For example, a recent survey reported that 14.1% of respondents had used cannabis within the previous 12 months—nearly double the 1994 figure of 7.4%. Did this represent a true increase in cannabis use or were respondents simply less afraid to admit their (illegal) use in 2004 than they had been in 1994? Based on the survey data alone, there is no way to know.

One method of handling this problem is to conduct a poll with a *randomized response*. For example, pollsters ask each survey respondent to first roll a die (in secret). If the die shows 6, they should simply say Yes to the next question. Otherwise, they should truthfully answer Yes or No. This way, respondents are less afraid to respond truthfully: even if they say Yes, pollsters will never know if they actually used cannabis, or if they were a non-user who just happened to roll 6.

How can pollsters make use of such a survey? By using probability theory! Suppose 12,000 people were interviewed, and were each asked to first roll a die and then respond as above. Suppose 3,800 of them replied Yes, and the rest replied No. Now, on average about one out of six people (2,000 people) will roll a 6, and will answer Yes. Subtracting them from the survey results, we are left with about 10,000 people who truthfully answered the question, of whom about 1,800 responded Yes. This result would indicate that about 18% of the respondents actually used cannabis within the past year. And this estimate (18%) would probably be quite accurate, since the randomized-response design allows people to answer honestly.

Finally, there are times when confusion results simply from how a poll is designed and interpreted, even if the results themselves are completely valid. For example, it was widely reported that in exit polls conducted during the 2004 U.S. presidential election, when voters were asked about the most important issue affecting their vote, more chose "moral values" than any other option. Critics of the eventual winner, George W. Bush, argued that this proved that Bush's supporters were all extremist Christian conservatives. Conservatives argued that the poll result indicated a fundamental shift of American public opinion toward a more religious and pious perspective. But a closer examination revealed that just 22% of respondents had selected "moral values." Furthermore, this phrase may have meant different things to different people, while the six other choices offered by the survey (things like "Iraq" and "terrorism" and "health care") were more specific. In fact, "terrorism" was chosen by 19% of voters, and "Iraq" by 15%, for a total of 34%. So, if the poll design had combined these two issues into a single category ("security" or "foreign policy"), the poll might have concluded instead that "security" was voters' number one concern—quite a different conclusion and probably a more accurate one.

Poll Bias

The single biggest potential problem with polls is bias.

We are all familiar with people who draw conclusions about "what everybody thinks" from asking a "sample" of people who are all close friends. From the Probability Perspective, we would say that their polling sample was biased: just because their friends agree with them doesn't mean everybody else does, too.

Similarly, people have a tendency to notice only those facts or arguments that support their position and to ignore those that don't. For example, a friend of mine who believes much more than I do in inherent behavioral differences between boys and girls noted that her young son enjoyed playing with toy trucks, which "proved" his innate masculine nature. Her son also really likes flowers, but my friend dismissed this interest as an insignificant and irrelevant aberration, thus reinforcing her belief in traditional gender stereotypes.

Advertisements are rife with examples of polling bias. For example, numerous television commercials advertise various exercise or diet plans to assist with weight loss. These commercials invariably include testimonials from satisfied customers who rave about how many pounds or dress sizes or inches they have lost using this product. The problem is that these customers have been selected by the company. There may well be many customers who did not lose weight (or even *gained* weight) using their product, but you won't see them on television. The testimonials shown form a *biased sample* and carry no statistical weight (as opposed to body weight) at all.

For similar reasons, we should never trust any poll conducted directly by a commercial enterprise or political party that has a stake in the results. Not only can selective reporting and biased sampling affect the results, but so can the phrasing of a question or tone of voice in which it is asked. Consider these two questions: "Do you agree that overbloated governments should reduce tax rates for our hard-working private sector, to allow it to be more efficient and to create more jobs, thus benefiting everyone?" and "Do you agree that wealthy multinational corporations should be allowed to keep even more of their huge profits, while contributing even less to society's basic needs such as health care, education, and public transit?" These two questions ask essentially the same thing, but they are likely to get very different answers.

It is fine for corporations or politicians or even eccentric individuals to commission and fund polls. However, the polling itself should always be done by an independent, professional, experienced polling company. Only if its operators can ensure that the poll subjects are chosen randomly from the population as a whole, without any bias or favoritism, can a poll's results be considered valid.

The Aging Skater

As a child, you always enjoyed the ice-skating outings organized at your school. You and your classmates crowded onto yellow buses and were driven to the local skating rink, where you zipped around the ice. You were never one of the best skaters, but you were certainly above average, and you always had a good time.

Now, years later, you see a sign for a weekly "Adult Skate" at a downtown rink. For old times' sake, you buy a pair of secondhand skates, and glide out onto the ice. Your new skates work fine, the old moves come back to you, and you're feeling pretty good.

But then you look around the rink and are shocked to see blistering speed, graceful turns, backward maneuvers, and even leaps and twirls, from most of the other skaters. Of the 100 or so skaters out on the ice, you are without doubt one of the worst.

How can this be? Has your skating really worsened that much? You don't think so. But if not, how could you have gone from being an above-average skater to one of the worst? What happened to all those kids who skated so much more poorly than you did?

And then it hits you: you are a victim of sampling bias. The only people who attend Adult Skate hour on a regular basis are the people who enjoy and excel at skating. And what about all those other people, the ones who couldn't skate very well as children? They rarely go skating anymore, so few of them are here today to make you look good by comparison.

We see that bias is a real problem with the sort of informal "polls" that we all conduct, implicitly, on a daily basis. But what about official, formal polls conducted by reputable polling companies? Aren't they careful enough to avoid bias in their polls?

Usually the answer is yes. Professional polling companies, with many years of experience, do indeed avoid injecting bias into their polls. That is why their predictions are usually very close to the actual results on election day. However, in certain cases, other forms of bias can enter.

For example, when a conservative government was elected in Ontario in 1995, it instituted an aggressive welfare-reform plan which, among other things, reduced welfare benefits by 21.6% and made the conditions to qualify for welfare more stringent. The government claimed these changes would encourage recipients to end their dependency and find jobs; critics claimed they would cause misery and hardship for society's neediest members. In fact, the number of people on welfare quickly decreased, but there was widespread disagreement about why this occurred, or what had become of the former welfare recipients.

In response, the government commissioned a poll. In October 1996, a private polling company telephoned people who had gone off welfare in May of that year. The company reported that 62% of these people listed a new job as their reason for no longer collecting welfare. "The vast majority are leaving the system for employment-related reasons," declared the minister of social services proudly.

But there was a problem. The polling company had attempted to contact all 16,219 people who had gone off welfare in May 1996, but had managed to track down only 2,100 of them. What happened to the other 14,119? Presumably many of them had been forced to move, or no longer had a phone, or were otherwise uncontactable. In short, the 2,100 respondents constituted a biased sample. It was likely that most of the 14,119 uncontacted people, in addition to 38% of the contacted people, had *not* found a job and were worse off than ever. However, it took several days of public debate—and a careful consideration of sampling bias—before this conclusion was clearly articulated.

Election-Night Jitters

A different form of bias can occur on election night. Often it takes several hours for all the ballots to be counted and reported, and the results come in to media centers in dribs and drabs. You might think that this is an unbiased way of getting information—after all, the election workers themselves are scrupulously neutral, and they are all trying to report the results as quickly as they can. If the delays don't have any relation to how people voted, they shouldn't affect how the results are reported. However, sometimes there *is* a connection between the votes and the delays.

The most spectacular example is the night of the Quebec sovereignty referendum of 1995. Much was at stake, the result was very close, and millions of Canadians were glued to their television sets that evening. The referendum voting was scheduled to end at 8:00 P.M., and by 8:30 results started trickling in. Early returns showed the Yes side in the lead. As more results came in, the Yes side continued to lead slightly. Even a couple of hours later, with the majority of the votes counted, the Yes side still had just over 50% of the votes cast. It wasn't until nearly 11:00 P.M. that the No side inched ahead, and almost 11:30 until the main national television network, the CBC, officially declared the No side the winner.

How could this be? How could the Yes side be in the lead with over half of the votes counted and yet go on to lose? If a poll of just 1,000 citizens is considered accurate, then shouldn't a sample of over half of the voters be even more so?

Part of the answer, of course, is that the vote was so close that any little change in the count could affect which side was ahead. But a more fundamental answer is that the delay in counting the votes was not purely random. Indeed, the biggest delays happened in the Montreal urban area. So, by around 10 P.M., most of the votes outside of Montreal had been counted, while a significant number of the votes within Montreal had not. Furthermore, Montreal, with its large non-francophone population and more global outlook, is home to a majority of No supporters. So, the partial vote count at 10 o'clock was a biased sample; it counted too many non-Montreal Yes votes compared with not enough Montreal No votes. This bias was just enough so that, even though overall a majority of Quebecers voted No, a majority of the votes counted by 10 o'clock were Yes. Such are the perils of biased sampling, even on election night.

Another interesting example was the reporting of the Florida results in the U.S. presidential election of 2000. These results are now remembered for protracted post-election arguments about butterfly ballots, hanging chads, manual recounts, partisan election officials, and Supreme Court decisions. But even before that debacle, the Florida results provided additional confusion. At approximately 10 o'clock EST on election night, most television networks predicted that Al Gore would handily carry Florida, and several networks even announced that Gore would therefore win the entire presidential election. They were forced to retract their predictions about an hour later, as the Florida results suddenly became "too close to call," which turned out to be an understatement.

What happened? The networks had forgotten about time zones. The majority of the state of Florida is in the Eastern time zone. However, the northwest tip of Florida (the Panhandle) is in the Central time zone. Since polls closed at 8 P.M. local time everywhere in the state, this meant that vote counting from the Panhandle began an hour later. And since Panhandle voters tend to vote mostly Republican, the networks underestimated Republican votes in Florida, and incorrectly predicted a solid Gore victory there. Indeed, the networks were lucky that the Florida results later became controversial due to the extreme closeness of the final tally; otherwise their error would have been more heavily criticized.

Bush versus Kerry

The 2004 U.S. presidential election that pitted George W. Bush against John Kerry also provides insights into various polling issues.

Interest in this election was extremely high, due to its expected closeness, and also to the polarization that Bush's first term as president had provoked (most people either loved him or hated him). Numerous pre-election polls were conducted. In the weeks before November 2, two firms ran polls every single day, and virtually every polling company and media outlet sponsored at least one poll.

The results of these pre-election polls were all found to be so close as to be "within the margin of error." Virtually every commentator simply declared that the race was "too close to call." One company, Rasmussen Reports, surveyed between 500 and 1,000 American citizens each day for many months, and even they determined that either candidate could win the election, since so many states were considered to be "toss ups." Knowing that an election will be very close is important information. It inspires party volunteers to campaign tirelessly, citizens to cast their ballots, and candidates to pay close attention to voters' opinions. Unfortunately, it doesn't help with election polls' primary function, which is to figure out who will win.

Despite the predicted closeness, the vast majority of pre-election polls showed Bush in the lead, ever so slightly, by just a few percentage points. Every one of those polls was closer than the margin of error, and thus a "statistical dead heat." However, it was possible to combine all of these different poll results together, to effectively create one single much larger poll.

How does this help? Well, as we will see in the next chapter, a much larger poll corresponds to a much smaller margin of error. If a whole bunch of polls all show Bush slightly ahead, then this is much more convincing than if just a single poll shows the same thing. So, several days before the election, it seemed clear (to me, anyway) that the polls, when combined, were predicting a narrow Bush victory. Although nearly everyone else was declaring the election "too close to call," I believed that, barring any major and unexpected last-minute opinion shifts, Bush would win by a few percentage points.

While most pre-election polls were pretty consistent (showing Bush ahead by a few percentage points), there were a few that weren't. In particular, one poll by *Time* magazine, and another by *Newsweek,* showed Bush ahead by a full 10%. Scott Rasmussen, who was himself conducting daily tracking polls, analyzed the difference and concluded that the magazines had sampled too high a fraction of registered Republicans. That is, those two polls had a sampling bias toward the Republican Party. When this bias was accounted for, the conclusions were similar to the other polls: Bush ahead by a few percentage points.

Further excitement came on election night. Going in, it was a virtual certainty that if Bush won Florida and Ohio, he would win the election. On the other hand, the voter turnout was higher than expected, which some thought would favor Kerry. Also, exit polls (where a sample of voters are asked how they voted, right after they leave the polling booth) seemed to show Kerry slightly ahead in Ohio, and tied in Florida. Could he pull off a last-minute upset?

When the actual vote-counting began, Bush quickly developed a lead of between 4% and 5% in both Florida and Ohio, contradicting the exit polls. Were the exit polls wrong, and Bush would win these states and the election? Or, was this a situation like the vote-counting during the Quebec referendum, where certain Republican-dominated regions were counting their ballots more quickly than everybody else and blurring the result?

The network analysts provided little assistance with this question. Even after 75% of the Florida vote had been counted, with Bush still up by about 52% to 47%, Jeff Greenfield of CNN declared that conclusions were premature because "we don't know where those votes are coming from." That was rather an odd statement, since a live, county-by-county breakdown of the Florida votes was publicly available on CNN's own Web site. Examining those counts, I saw that the vote-counting was proceeding fairly evenly: some counties were done counting, and others were only about halfway through, but there was no clear pattern of pro-

Bush regions finishing their counting faster, or pro-Kerry regions finishing their counting more slowly. There was no reason to think the result would change much with the later vote-counting. Within an hour and a half of the Florida polls closing, I knew that Bush had won Florida.

The case of Ohio was more interesting. Early vote-counting there also showed Bush ahead by about 52% to 47%. However, due to large voter turnout—and, one suspects, inadequate election infrastructure—many Ohio voters had to wait in line for four hours or more, and vote-counting was proceeding very slowly there. Several hours after the polls closed, still only about one third of the votes had been counted. Would Bush's lead hold up? A closer examination of the numbers (again from the CNN Web site) showed that one particular Ohio county, Cuyahoga (Cleveland and surroundings), was of particular interest. It contained a huge number of voters (well over half a million, compared with just 20,000 in many other Ohio counties). Furthermore, its votes were going to Kerry by a two-to-one margin, and less than half of its votes had been counted. Kerry would surely gain votes as the rest of Cuyahoga's results came in. However, a quick calculation convinced me that Kerry's Cuyahoga County gain would amount to less than half of what he needed to overcome Bush's lead.

So, by 10 P.M. Eastern time, I was sure (though not pleased) that Bush would win both Florida and Ohio, and thus the election. Meanwhile, most of the television networks, spooked by their erroneous predictions four years earlier, didn't come to any conclusions about Florida until very late that night, nor about Ohio until late the next morning.

What was the final result? Bush won the popular vote by about 51.1% to 48.0%, very close to the average of the pre-election polls. He won Florida by 52.1% to 47.1%, just as the early counting had indicated. And he won Ohio by 51.0% to 48.5%, a somewhat smaller margin than in the early vote-counting (due to Cuyahoga) but still a clear victory. In short, despite the network's caution, the election result really showed very few surprises given the pre-election polls and initial vote-counts.

The only remaining puzzle is those exit polls. The old Voter News Service had been dissolved in favor of a new National Election Pool, supported by six leading media outlets (ABC, CBS, NBC, CNN, Fox News, and Associated Press). They conducted extensive exit polls, using all the latest techniques. Why did these exit polls show a lead for Kerry in Ohio, and such a close race in Florida?

Well, not all voters are willing to talk to pollsters after casting their ballots; some are too busy, or in too much of a hurry, or prefer to keep their votes private. Apparently, in the 2004 presidential election, Kerry supporters—angry at Bush and proud of it—were more willing to talk to pollsters than Bush supporters were. So, the exit polls (unlike the other pre-election polls) showed a higher percentage of support for Kerry than he really had: in short, the exit polls were biased.

Polls are extremely important and influential, but they must be interpreted properly. Biased or misleading polls are worse than useless. Even high-quality polls cannot predict the future, nor completely overcome dishonest responses. But at least they can give us a snapshot of current opinions and a hint of what might lie ahead.

11

Nineteen Times Out of Twenty

Margins of Error

In the previous chapter, we saw that polls have many limitations. They do provide a useful snapshot, and even allow for communication and collaboration among citizens. On the other hand, they cannot account for future changes. They can be biased if they are not performed properly, or if certain types of respondents participate at lower rates. They are vulnerable to misleading responses or to certain categories of citizens voting in higher or lower proportions. Indeed, all that a poll's reported "margin of error" really measures is how far off the poll's results are likely to be from the result that they would have gotten if they surveyed the entire population instead of merely a sample.

Even so, this margin of error is still an important quantity. If only a few people are surveyed, the results will not be very useful no matter how professionally the poll is conducted. The more people who are surveyed, the more likely it is that the results will be close to the responses of the population as a whole. But how likely? How close? How is a margin of error such as "accurate within 1.4 percentage points, 19 times out of 20" actually calculated?

Flipping Coins

From a probability perspective, surveying citizens is similar to counting the number of heads when flipping coins. The main difference is that with coins we know in advance that heads has a probability of 50%, but with polls we don't know in advance what fraction of citizens support a particular position. Indeed, that is the difference in a nutshell between probability theory and statistical inference: in probability theory we know the individual probabilities in advance, while in statistical inference we don't.

So, to understand margins of error, imagine flipping coins, without knowing in advance that the probability of heads is 50%. Suppose you flip many coins. How close will the fraction of heads that you observe be to the true probability of heads, which is 50%?

If you flip just one coin, it will be either heads or tails. Thus, your fraction of heads will be either 100% or 0%, neither of which is close to the figure of 50%.

If you flip two coins, then the probability is 25% that you will get 100% heads, and 50% that you will get 50% heads (one heads out of two flips), and 25% that you will get 0% heads (both flips will come up tails). We can plot these probabilities on a graph as shown in Figure 11.1.

Figure 11.1 Probabilities for Percentage of Heads When Flipping Two Coins

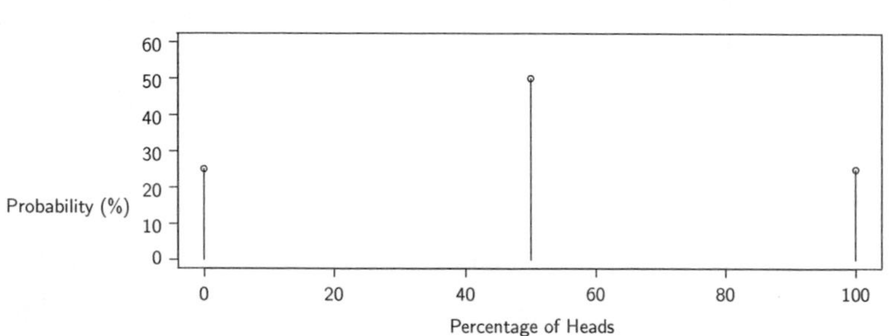

Again, this is not a very reliable way of getting close to the true answer of 50%.

On the other hand, if you flip 10 coins, there is less than one chance in 1,000 that you will get 100% heads, and similarly less than one chance in 1,000 that you will get 0% heads. In fact, the probability of getting exactly 50% heads is 25%, while the probability of getting either 40% or 60% heads is 21% each. We can plot all of these probabilities on another graph:

Figure 11.2 Probabilities for Percentage of Heads When Flipping 10 Coins

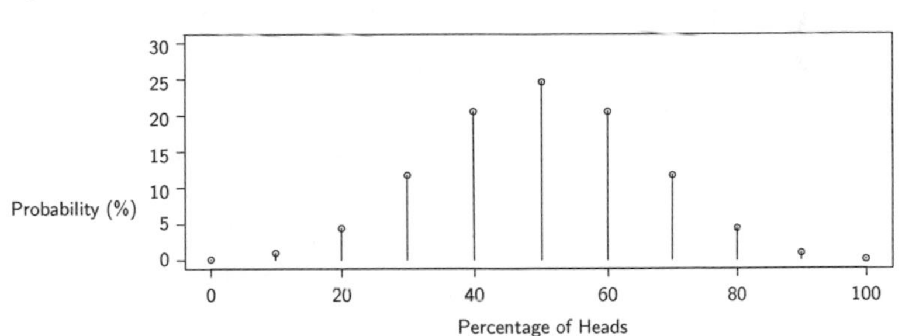

What Figure 11.2 shows is that, when flipping 10 coins, you are most likely to get 50% heads, but you're also fairly likely to get 40% or 60% heads, and somewhat likely to get 30% or 70% heads. On the other hand, you are very unlikely to get 0% or 10% or 90% or 100% heads.

So how do we convert these results into a margin of error? We have to combine the probabilities for various possible outcomes, to get a total probability of at least 95%, or "19 times out of 20." Indeed, if we add up the probabilities of getting 20% heads or 30% heads or 40% heads or 50% heads or 60% heads or 70% heads or 80% heads, we get a total probability of 97.9%, which is above 95%. Thus, combining all these probabilities together gives us a range of heads percentages that will occur at least 19 times out of 20:

Figure 11.3 95% Range for Percentage of Heads When Flipping 10 Coins

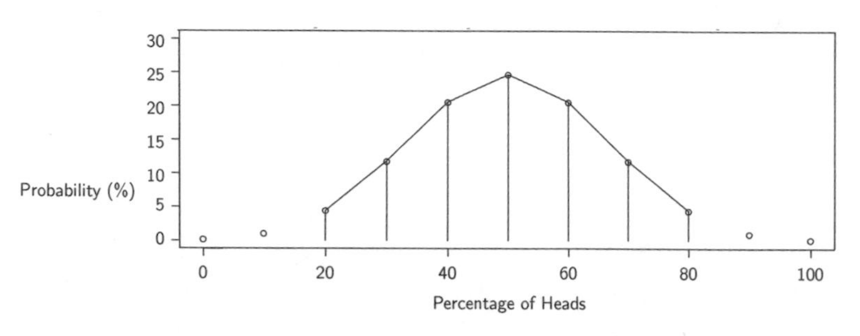

Suppose you flip 10 coins as an experiment, and you repeat this experiment 20 times. What will happen? You will get between 20% and 80% heads, approximately 19 times out of 20. That is, 19 times out of 20 you will be within 30% of the true (50%) probability of heads. Thus, flipping 10 coins has a margin of error of 30%.

We can translate our understanding about flipping coins into an understanding of polls. If you poll 10 citizens, your margin of error will be 30%, just as it is for coins. That is, 19 times out of 20, the percentage of those polled who support your candidate will be within 30% of your candidate's true support among the whole population. To put it another way, your 10-person poll results will be accurate to within 30%, 19 times out of 20.

A 30% margin of error is much too large. You might think your candidate has 65% support, when she really has just 35% support, a huge discrepancy. To improve matters, you have to survey more people (or flip more coins). Indeed, the Law of Large Numbers tells us that the more coins you flip, the more likely it is that the percentage of heads will be close to 50%.

If you flip 100 coins and compute a 95% range as before, the resulting graph looks like Figure 11.4.

Figure 11.4 95% Range for Percentage of Heads When Flipping 100 Coins

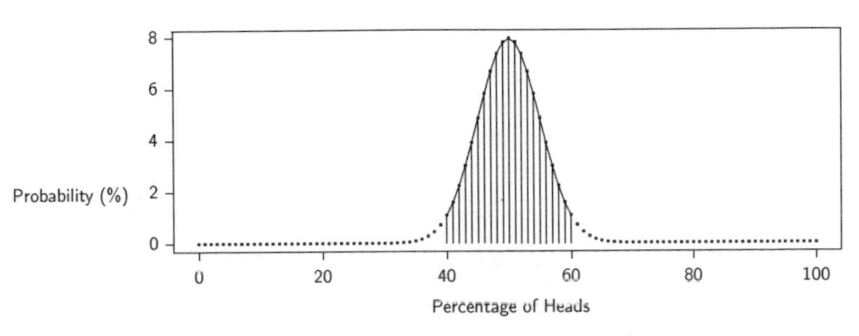

That is, when you flip 100 coins, you will get between 40% and 60% heads, 19 times out of 20. (If you don't believe me, try it.) Thus, the margin of error is now 10%, much smaller than the 30% margin of error we had earlier. So, if you poll 100 citizens, your margin of error will be 10%; that is, 19 times out of 20, you will be within 10% of the true level of support.

Using the Bell Curve

What if you flip 1,000 coins? Or 10,000 coins? Do you need to keep graphing the results and adding up probabilities each time?

Fortunately, no. We can use the old mathematical trick of *finding a pattern*. Way back in 1733, a French Huguenot named Abraham de Moivre was the first to notice that the shapes of these graphs get closer and closer to what we now call the *bell curve*.

Figure 11.5 The Bell Curve

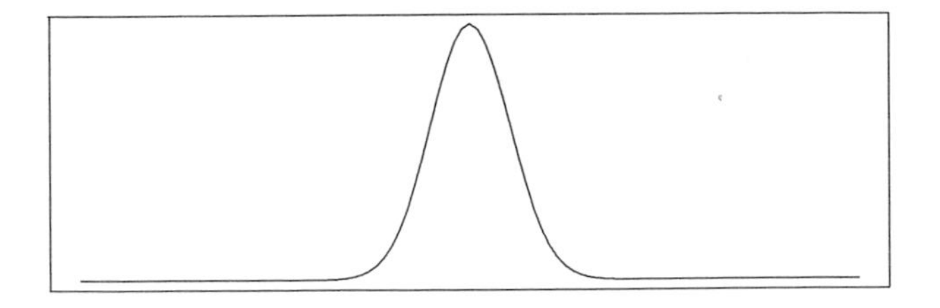

This convergence to the bell curve is called the *central limit theorem*. It was later studied in more detail by the French mathematician Pierre-Simon Laplace and the great German mathematician Johann Carl Friedrich Gauss. They showed that the bell curve—also called the *normal* (or *Gaussian*) *distribution*—arises from many different probability experiments, not just flipping coins. (And the approximation gets more accurate the more coins you flip. If you flip 100 coins, the results look a lot like the bell curve, but if you flip just 10 coins they don't look so close.)

Using this bell curve, it is a simple matter to determine a formula for the margin of error when flipping many coins. All we must do is measure an area under the bell curve that equals a total of 95% of probability. (The modern way to do such measuring is by performing numerical integration on a high-speed computer, but it can also be done with a ruler and pencil if you are careful and patient.) For a "standard-sized" bell curve, the 95% region is between −196% and +196%:

Figure 11.6 95% Margin of Error for the Standard-Sized Bell Curve

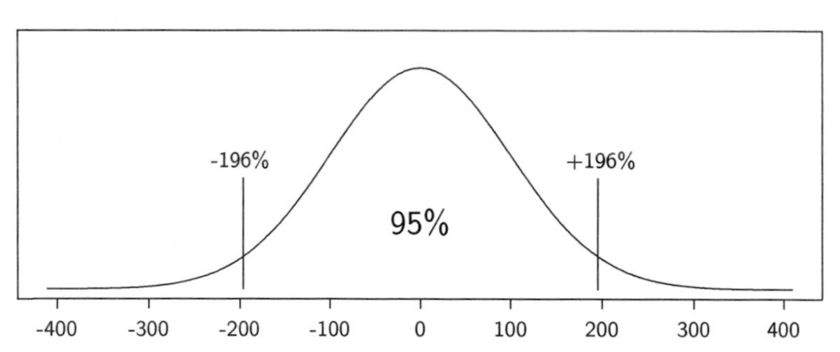

The probabilities when you flip lots of coins are just like the standard-sized bell curve, except narrower by a factor of two times the square root of the number of coins, corresponding to the "standard deviation." What this means is that when you flip lots of coins, the margin of error is just like that for a standard-sized bell curve, except divided by two times the square root of the number of coins.

Since 196% divided by 2 equals 98%, this provides a simple rule: when you flip lots of coins, your margin of error is equal to 98% divided by the square root of the number of coins flipped. That is, 19 times out of 20, your fraction of heads will differ from the true probability of 50% by no more than 98% divided by the square root of the number of coins you flipped.

For example, suppose you flip 100 coins. The square root of 100 is 10 (since 10 times 10 equals 100), and 98% divided by 10 is equal to 9.8%. So, if you flip 100 coins, your margin of error will be approximately 9.8%. This is very close to the 10% margin of error we found earlier.

Suppose now that you flip 1,000 coins. If we calculate 98% divided by the square root of 1,000 using a calculator, we find that it is equal to 3.099032%, or about 3.1%. So, if you flip 1,000 coins, your margin of error will be about 3.1%. This means that if you flip 1,000 coins, 95% of the time, or 19 times out of 20, you will get between 46.9% and 53.1% heads.

So What About Those Polls?

We now have a simple rule for margin of error when flipping coins: divide 98% by the square root of the number of coins flipped. Fortunately, the same formula works for polls. That is, to determine a margin of error for a poll, you can simply take 98% and divide it by the square root of the number of people surveyed. It really is that simple.

Four polls were described at the beginning of the previous chapter. How do their asserted margins of error mesh with this newfound formula?

The Spanish poll surveyed 24,000 citizens and claimed a margin of error of 0.64%. If we compute 98% divided by the square root of 24,000, we obtain the figure 0.6325873%, which rounds up to 0.64%.

The Canadian poll surveyed 5,254 citizens, and asserted a margin of error of 1.4%. Indeed, 98% divided by the square root of 5,254 equals 1.352%, which rounds up to 1.4%.

The Australian poll surveyed 1,397 citizens, and claimed a margin of error of plus or minus 2.6 percentage points. And indeed, 98% divided by the square root of 1,397 is 2.622%, very close to 2.6%.

The U.S. poll surveyed 1,212 citizens, and asserted a margin of error of plus or minus 2.9 percentage points. And indeed, 98% divided by the square root of 1,212 is 2.815%, which they kindly rounded up to 2.9%.

So there you have it. All those fancy claims about margins of error and accuracy 19 times out of 20 amount to just dividing 98% by the square root of the number of people surveyed. Now that you know this, you can compute margins of error of polls all by yourself. (Sometimes polling companies will use more sophisticated analyses, to derive somewhat smaller margins of error. This is especially relevant when a proposition or party has support very close to 0% or to 100%. But most of the time, companies simply divide 98% by the square root of the number of people surveyed, just as you can do yourself.)

To obtain an accuracy level (or "confidence level") of 99 times out of 100, instead of just 19 times out of 20, you need a slightly larger area under the bell curve. Specifically, you have to replace 98% with 129%, so now you divide 129% by the square root of the number of people surveyed. With 1,397 people surveyed (as in the Australian poll), this 99%-confident margin of error works out to 3.5% (instead of 2.6%). So, in that Australian poll, in addition to saying that their results are "accurate within 2.6 percentage points, 19 times out of 20," they could also claim that their results are "accurate within 3.5 percentage points, 99 times out of 100." Both of these are true statements.

If you survey more people, the margin of error gets smaller. Indeed, as you survey more and more people (or flip more and more coins), the randomness in your sample tends to cancel out, so the uncertainty is reduced. In colloquial terms, the more people you ask, the more you know. That is just common sense. But with our new formula for margin of error, we can quantify this truism in precise mathematical terms.

On the other hand, the poll's margin of error is different from its accuracy. Since those four elections have all passed, we can judge each of the four polls on the basis of how closely they predicted the results.

The Australian poll was very accurate. They predicted the incumbent prime minister to lead the opposition by 52% to 48%, which was extremely close to the final result of 52.8% to 47.2%. No argument there.

The U.S. poll also did well. It predicted George W. Bush with a lead of 47% to 45% over John Kerry. In the actual election, Bush won the popular vote by about 51.1% to 48.0%. This indicates that votes for other candidates (besides Bush and Kerry) mostly collapsed, and were significantly less than predicted. But on the important matter, namely the relative strength of Bush versus Kerry, the poll provided a very accurate election forecast.

The Canadian poll, on the other hand, predicted 32.6% support for the Liberals, 31.8% for the Conservatives, and 19.0% for the NDP. The actual election result was Liberals 36.7%, Conservatives 29.6%, and NDP 15.7% (with the rest of the votes going to other parties). So the poll's predictions were rather poor, and indeed outside their margin of error. As discussed earlier, a last-minute shift of voter intentions was to blame for this. Polls still have their limitations, even after the margins of error are carefully computed.

Finally, the Spanish poll was the least accurate of all. It had predicted 42.2% of the vote for the governing Popular Party, and 35.5% for the opposition Socialist Party. The final result was almost exactly the opposite: 43.27% for the Socialist Party and 37.81% for the Popular Party. These results indicate the last-minute shift in public opinion caused by the terrorist train bombings just before the election, and by the perceived inadequacy of the governing party's response. Once again, actual events and changing opinions proved too much for the polling companies, margins of error notwithstanding.

How Close Is Close?

Margins of error also provide a new perspective on the closeness of elections.

Some elections are a landslide, but many elections are fairly close. For example, in both the Australian and American elections of 2004, the margin of victory was just a few percentage points—close enough that, even on election day, the outcome was still somewhat in doubt (making the election returns worth watching). On the other hand, every once in a while there is a vote that is so close that it defies our every expectation. Two examples leap to mind: the 1995 Quebec sovereignty referendum and the 2000 U.S. presidential election.

When the vote-counting for the 1995 referendum was finally done, there were 2,308,360 Yes votes, and 2,362,648 No votes (plus 85,501 spoiled ballots). Of the valid ballots, the vote was 49.42% Yes versus 50.58% No. The difference was 54,288 votes, few enough people to fit comfortably into a football stadium. Clearly, this was a close vote. But how close? Compared with what? One mode of comparison is to pretend that every voting Quebecer simply flipped a coin and voted Yes if the coin was heads, No if it was tails. If they *had* done so, how close would the vote have been?

This is a question we know how to answer. If each of those 4,671,008 valid votes was instead decided by flipping a coin, then on average there would be 50% heads. However, the *margin of error* would be equal to 98% divided by the square root of 4,671,008, or 0.045% (equivalent to just 2,102 votes). This means that, 19 times out of 20, the winning side would have gotten less than 50.045% of the vote. In fact, the winning (No) side got 50.58%, which is significantly more. So, although this Quebec referendum result was extremely close, it was still far less close than the "margin of error" from a true 50–50 vote. Had Quebecers voted by flipping a coin, the result would have been far closer.

This brings us to the 2000 U.S. presidential election. There, the popular vote was actually won by Al Gore (although he lost the election itself, due to the Electoral College election system). He received a total of 50,999,897 votes, compared with 50,456,002 for George W. Bush. Thus, of the 101,455,899 votes cast for either candidate (ignoring votes for other candidates), Gore received 50.27% of the vote compared with just 49.73% for Bush. This is very close, but again outside the corresponding margin of error (which would be 98% divided by the square root of 101,455,899, or 0.0097%, corresponding to a victory of just

50.0097% to 49.9903%). So, in terms of popular vote alone, Gore was the undisputed winner.

However, the real interest in this election was the Florida count. Whichever candidate won the most votes in this state would obtain enough Electoral College votes to win the election and the presidency (regardless of the nationwide popular vote). The declared final Florida vote tally (after much controversy about recounts and hanging chads and butterfly ballots and Supreme Court decisions and so on) was 2,912,790 votes for George W. Bush and 2,912,253 votes for Al Gore, for a difference of just 537 votes. Of the 5,825,043 votes cast for either candidate (again ignoring votes for other candidates), this corresponds to 50.0046% of the vote for Bush, and 49.9954% of the vote for Gore. Even accepting this count as accurate (which it may not have been), it is exceptionally close. Just how close?

Suppose each Florida voter had instead flipped a coin and voted for Bush if it was heads, or Gore if it was tails. In this case, the margin of error would have been 98% divided by the square root of 5,825,043, or 0.04%. Thus, 19 times out of 20, such a coin-flipping vote would produce a result of 50.04% to 49.96% or closer, corresponding to a difference of 4,730 votes. Such a result would indeed be very close. However, the actual result was nearly nine times closer.

The Florida returns were far closer—about nine times closer—than we could ever expect them to be, even for an equally divided electorate, even if each voter was simply flipping a coin. The result was simply outside the norm of even an exceptionally close election.

In the days before the 2004 U.S. election, there was much discussion of whether the result would be as close, and as confusing, as that of 2000. However, I doubted that it would be. No matter how closely divided the electorate is, the margin of error alone is simply too large to produce such a narrow victory with any regularity.

On the other hand, out of *all* of the different U.S. election contests for senators, representatives, governors, and so on in all 50 states, it was still likely that some other race, somewhere, would be extremely close. Indeed, the 2004 election for governor of the state of Washington had, after three recounts, a margin of victory for Democrat Christine Gregoire over Republican Dino Rossi of 1,373,361 votes to 1,373,232—a 129-vote difference, less than *one twenty-fifth* as large as the 3,248-vote difference corresponding to the margin of error. With so many different electoral races to choose from, there will always be a surprise somewhere or other.

Errors and uncertainty are with us on a daily basis. Often we cannot compute the margin of error with a formula as precise as "98% divided by the square root of the number of respondents," but a general understanding of the meaning of "margin of error" can still be very helpful.

An Uncertain Day

You drag yourself out of bed and get ready for work. The bus is supposed to pass by at 8:15 A.M., but last week it was 10 minutes late on Monday and Thursday, and 10 minutes early on Tuesday and Friday. The bus schedule clearly has a large margin of error. To be on the safe side, you go to the bus stop at 8 o'clock.

At work, you get a memo from your boss saying that she will come to your office at 10:30 to pick up your accounting report. Your boss is very punctual, so the margin of error for her arrival time is virtually zero. You finish your report at 10:29, and have it waiting for her upon her arrival at precisely 10:30.

At noon you go to lunch with a few colleagues, although you have to be back in the office for a meeting at 1 o'clock. You consider going to Billy's Deli, which usually has very quick service. However, a few times Billy has been backed up, and it has taken half an hour or more to get served. Clearly Billy has a large margin of error. Instead, you eat at Clare's Coffee Shop, where every order gets filled in exactly 10 minutes, with hardly any margin of error at all.

In the afternoon, you have to reformat your report according to your boss's instructions. With some careful thought, you figure out just how this should be done. You consider asking your assistant to carry out your plan. However, while she sometimes follows your instructions precisely, she sometimes gets overly "creative" and does a poor job; her margin of error is too high. You decide to do the reformatting yourself.

That evening, safely home again, you open your refrigerator. You reach for a carton of milk, but see that it is two days past its expiration date. Since milk only lasts for a few weeks anyway, there isn't too much margin of error there. You decide to play it safe, and pour the milk down the drain.

Instead, you reach for some orange juice, and see that it is also two days past its expiration date. On the other hand, most juice lasts for several months, so its expiration date has quite a large margin of error. Two days before or two days after probably amounts to the same thing. You decide to drink the juice, and it is fine.

Finally it is time for bed. Your neighbors aren't making any noise at the moment. But sometimes in the middle of the night, they come back from a party and have a loud conversation, or they watch a late-night movie with the volume turned up, or their baby cries like an opera singer. Usually not, but the margin of error is just too great. You carefully switch on a fan, to drown out any potential noise and allow you a good night's sleep.

Margins of error provide a fresh perspective on life's uncertainty, and give us a measure of the reliability of what we observe. Often we are hoping to minimize the margin of error by taking more samples or observing more carefully. But we will see in the next chapter that uncertainty can sometimes benefit us, too.

12

Randomness to the Rescue

When Uncertainty Is Your Friend

In the movie *The Sting,* a friend worries that the Robert Redford character, Johnny Hooker, will be the mob's next murder victim. Attempting to prevent this horrible fate, the friend advises Hooker to avoid locations where the mob might find him. He wants Hooker to go somewhere so unexpected, so unpredictable, that even the mobsters, with all their hit men and weapons and connections, will be stumped. He says, "Don't go home tonight. Don't go to your usual bar. Don't go anyplace you normally go." He might have added, "Go someplace random."

We are all familiar with the notion of doing something unexpected (or random), to fool someone or to win a game or to entertain. Randomness brings much misfortune and misery, and we often consider it something to fear or avoid. But randomness can be very helpful, too. We make use of it every time we park our car illegally, "just for a minute," since it is unlikely that a police officer will happen by just then. And it's in the back of our minds every time we make backup copies of a computer file, since it is improbable that two or three different disk drives will all fail at once (and the more backups, the less likely that they will all fail). In many ways randomness is the enemy but, in other respects, randomness is our friend.

Randomness and Individuality

I just rolled an ordinary six-sided die 50 times in a row and obtained the following sequence of numbers:

34663463621632143466334452434226433642631152454334

Are you impressed? Probably not; anyone can roll a die 50 times. But perhaps you should be. You see, I have just created a sequence that no one has ever created before, and that the CIA couldn't reproduce in a million years!

It's true. In all of human history, there have been about 100 billion people. Even if each of them created similar 50-digit sequences, once per minute, for 100 years each, the odds would still be less than one chance in a 1 followed by 20 zeros that any one of them, from Julius Caesar to my neighborhood butcher, had ever created my special sequence above. In other words, it is absolutely inconceivable that this ever happened. My sequence has never appeared in any computer file, or on any Web page, or anywhere at all.

Suppose the CIA has a million computers, and they can each produce a billion sequences a second. It would take them 25 trillion years, working around the clock, to have even a 1% chance of hitting on my special sequence. In other words, it will never happen. In two minutes, sitting in my bedroom with a Monopoly game cast-off die, I have stumped the entire worldwide resources of the CIA.

All the king's horses
And all the king's men
Will never produce
That sequence again

What power. The power of randomness. The next time you are sitting with a die, think of the power in your hand.

The reason for this power is simple: each die can come up six different ways. So, the total number of possible 50-digit sequences is equal to 6 multiplied by itself 50 times, which is approximately equal to the number written as 8 followed by 38 zeros—an incredibly huge number. And each of those many sequences has the same probability, making it virtually impossible to guess which sequence has occurred.

There is an old mathematicians' tale (first imagined by the French mathematician Émile Borel in 1913) about letting a million monkeys each hit typewriter keys at random until the end of time, and how after typing loads of garbage they would eventually reproduce all the great works of literature, purely by chance. This is indeed true, if the monkeys had an *infinite* amount of time available to them. But we see just how impracticable this scheme really is. It would take those monkeys over a trillion, trillion, trillion, trillion years to have even a 1% chance of typing the sequence "It was the best of times, it was the worst of times." Human authors shouldn't drop their pens just yet.

You might wonder, do we really need to apply randomness? Couldn't I instead have just written down whatever 50 digits came to mind, not bothering to roll a die at all? Perhaps, but perhaps not. When people just try to make up a random sequence, they inevitably follow certain patterns. Perhaps they never repeat the same digit twice in a row. Perhaps they repeat certain digits too often. Perhaps they overuse one particular digit. Or perhaps they try too hard to use all digits equally. (For example, my sequence above has thirteen 3's but only four 1's, which is perfectly normal in a random sequence, but is unlikely in a made-up sequence.) So if you try to make up a sequence, you might accidentally choose something that someone else has already chosen. Or that the CIA might guess. Only randomness can guarantee that your sequence is unique.

A fun way to test for uniqueness is to use an Internet search engine like Google. If you do a Web search for a short random sequence of numbers or letters, like "axzqy" or "325794," you will doubtless find some Web pages that have that same sequence, just by chance. But if you search for a slightly longer sequence, like "axzqytuvb" or "3257948394," you will probably not get any hits. Out of all of the hundreds of billions of Web pages indexed by Google, not one of them contains your own personal random sequence. That's individuality.

And it's not just sequences. Consider all the choices you make in a random way every day, from what you eat for breakfast, to which shoes you wear, to what you say to the subway attendant on your way to work. Randomness shows us that we all do unique things all the time, and that even our daily trivialities involve unique choices that no one has ever made before.

Randomized Strategies

A popular children's game is called "Rock Paper Scissors." On the count of three, two players each put out one hand either in a fist (Rock), or flat (Paper), or with two fingers extended (Scissors). The winner is determined by the principle that Rock smashes Scissors, Scissors cut Paper, and Paper covers Rock. If both players choose the same option, no one wins and the game starts again. This game is sometimes used to decide contested issues, such as who gets to "go first" or who receives the bigger piece of pie.

Surprisingly, there are some people who take this game quite seriously. The Rock Paper Scissors International World Championships are held each year, and an official RPS champion is crowned.

If ever an RPS champion were to challenge me to a match, I would be appropriately nervous. Surely, with all of his RPS experience, the champion would excel at all of the psychology and guesswork required to intuit my next move. He would probably be an expert at discovering his opponent's patterns, such as always doing Scissors after doing Rock, or always choosing the option that would have defeated his previous choice. These are the kinds of patterns that we all (including me) tend to fall into, and ones that an RPS champion could use to his advantage.

Nevertheless, I could still arrange that, in the long run, I would defeat the champion as often as he would defeat me. I would do this by making my choices randomly, to avoid any patterns or predictability. Specifically, I would roll a die before each selection (the official RPS rules do not prohibit this). I would then choose Rock if the die showed 1 or 2, Paper if the die showed 3 or 4, and Scissors if the die showed 5 or 6.

My use of randomness would guarantee that no one could, in any way, find a pattern or predict my next move (provided the die was properly balanced and was hidden from my opponent's view). As a result, I would be guaranteed to win half the time (on average), no matter how clever or perceptive my opponent was, no matter what strategy he employed, and no matter how predictable my usual inclinations might be.

This perfect strategy, which cannot be defeated by any opponent, is related to the theory of *Nash equilibria,* which is named for its inventor, the brilliant but psychologically troubled mathematician John Nash (played by Russell Crowe in the movie *A Beautiful Mind*). Nash equilibria are choices of strategies such that no opponent can do anything to improve their fate. The mathematical subject of *game theory* proves that for any game (or other competition) in which each side has a finite number of choices to make (such as Rock–Paper–Scissors), you can always find such a Nash equilibrium, and thus prevent your opponents from outmaneuvering you. However, you may need randomness (like rolling a die) to achieve it.

Without randomness, there may be no way to avoid being outsmarted by an RPS champion. But with randomness, there always is.

Winning the World Series

It's the moment you've always dreamed of. The World Series. The seventh game. The bottom of the ninth inning. Two out. Bases loaded. Full count. Your team is ahead by one run. And you have just been sent in as a relief pitcher, to save the game for your team.

Thousands of fans cheer wildly as you trot out to the pitcher's mound. And now it all comes down to this, your next pitch. A strike and you win. A hit and they win. You stare into the batter's eyes, ready for the confrontation. Your two best pitches are your fastball and your slider. But which to use? You think long and hard; there won't be any second guesses here.

Meanwhile, the opposing team's dugout is a hive of activity. With all their computer equipment, mathematical models, and expert analysis, they have pulled up every bit of information they know about you. Your pitching record all the way back to Little League. Your pitching style and patterns. Your likes and dislikes. Your moods. They feverishly signal to the batter, sending secret messages you can't decode. Then the batter gives you a wink. He is anticipating. He knows!

You try to remain calm. You figure that if the batter guesses wrong, he will almost certainly strike out. However, if the batter guesses right and anticipates your pitch correctly, you figure he has a good chance of hitting it right out of the park. Now it becomes a guessing game. Can the opponents outsmart you, or can you outsmart them? Unfortunately, all your baseball training over the years hasn't left much time for mathematical models or battles of wits.

Don't give up so easily, you tell yourself. Probably they think you will throw a slider. After all, that's the pitch you used with a full count in the ninth inning in the game against Cleveland last month. (Or was that the month before?) So perhaps you should throw a fastball instead, to fool him. Or is that what they're expecting you to do? Or perhaps they have determined that you tend to alternate your pitches, so this time they're expecting a fastball and you should throw a slider. Or perhaps that's what they think you will think, so you should trick them by throwing a fastball after all.

It all seems so complicated. The pressure is getting to you. Your knees feel weak. Your palms start to sweat. You'll never be able to pitch like this!

While drying your hand on your baseball uniform, you feel a small bump—that coin your crazy uncle gave you earlier that day. "Take it to the game," he insisted. "It'll bring you luck."

Suddenly you get an idea. Why not let the coin make the decision for you? Yeah, that's an idea! With renewed confidence, you pull out the coin. "Heads fastball, tails slider," you think. You flip the coin, catch it against your baseball glove, and peek under your hand (being careful to hide the coin from the ever-present television cameras). Heads it is.

With renewed confidence, you stand up tall, wind up, and throw a beautiful fastball right past the surprised batter. Strike three!

It may seem crazy, or even cowardly, to use a coin toss to determine a baseball pitch. But in fact, this trick—under the more formal name of "randomized strategies"—is used all the time. Randomization is used to decide which manufactured parts to inspect, which employees to monitor, and which citizens to poll, among other things.

In the World Series example above, if you didn't use the coin, then maybe, just maybe, your opponents (with their sophisticated computer models and psychology) would have outsmarted you and guessed the pitch correctly. If so, they would have had a pretty good chance of hitting a home run and winning the World Series, let's say a 60% chance. So, while you may be skeptical that their computer models could predict so well, there is at least the possibility that they could.

On the other hand, with your randomized strategy, there is no way for your opponents to guess your pitch with confidence, no matter how clever they are and no matter how much they know about you. Rather, they will have just a 50% chance of guessing correctly, no matter what. This means that half the time they will guess correctly, and have a 60% chance of a hit; while the other half the time they will guess incorrectly, and always strike out. Thus, overall, they have just half of a 60% chance of getting a hit and winning the game.

So, with the randomized strategy, you have rendered useless all of your opponents' knowledge and insight and computer modelling, and given them just a 30% chance of victory, no matter what strategy they employ. Using your clever coin flip, you have guaranteed yourself a 70% chance of victory! Once again, probability saves the day.

Randomness and the Internet

Randomized strategies may be useful in sports, but do they arise in everyday life? Well, you use one every time you make a purchase over the Internet.

Computers seem by their very nature to be cold, logical, precise machines. They respond to precise commands; if you do the same thing twice, then they will do the same thing twice; if you do something different the second time, they will, too. (Of course, computers break down in seemingly random ways, but that's another story.) So it may come as a surprise to hear that computers in general, and the Internet in particular, could not get by without randomness.

Computers often need to make what are called "secure connections" on the Internet, for example to transmit your credit card number to an on-line shopping Web site, without allowing computer hackers to intercept your message. They do this through the use of randomly coded messages.

Nighttime Escape

Tonight's the night: you're going to sneak out after bedtime and go to the treehouse with your boyfriend. As he is leaving your house, you start discussing the plans with him. But just then your mother walks in.

This is terrible, you can't finalize the plans with your mother listening. And yet, without proper planning, you'll never be able to pull this meeting off. What to do?

After a moment's panic, the solution comes to you: a secret code. So, while uttering banalities about how you will see your boyfriend at school tomorrow, and what a gentleman he is, you casually hold up all 10 fingers ("Come at 10 o'clock . . ."), put one finger to your lips (". . . without making any noise . . ."), then move your hands up and down in a straight line (". . . and bring a ladder").

As your boyfriend leaves, you both smile, knowing your plans are all set. Unfortunately, your mother is also smiling. She has noticed your hand movements, too, and she is just as smart as your boyfriend. You watch broken-hearted as she padlocks your bedroom window.

Tricking your mother is hard enough. But what about foiling hackers when you use your credit card on the Internet? If your signal is intercepted, a computer hacker (or an international electronic spy agency like Echelon) might receive all the information that you send about your purchase. And, like your mother, they might understand this information just as well as the on-line store does. How can we get around this problem?

We need a way to transmit information that the receiver (like your boyfriend) can understand, but that the interceptor (like your mother) cannot. Secure computer connections rely on a protocol called "public key cryptography." This requires each computer to pick a random, secret key to facilitate the secure communication (for example, 128-bit security requires the equivalent of flipping 128 coins). The computers then use the theory of prime numbers to communicate in such a way that, by each using their own secret key, they can decode each other's messages. A third-party computer hacker, however, will not be able to break the code, even if they manage to intercept the entire conversation.

A crucial step in this secure connection is the selection of each computer's random, secret key. One option is for the shopper and the company to each flip a bunch of coins every time a purchase is made, and then type into their computers a long sequence of heads and tails. This would work fine, but it isn't very practical. Instead, your computer automatically picks a random key for you each time, using its built-in random-number generator. Such secure connections aren't just used for credit card numbers. Sophisticated computer users routinely connect to far-away computers using a "secure shell" to avoid the interception of passwords and other confidential information, and randomness is required each time.

The Internet is alive with randomness. Suppose two different people send you e-mail at precisely the same instant. How does the mail server computer make sure both messages arrive? It simply tells both e-mail messages (or, more precisely, all competing ethernet packets) to get lost. The messages are then assigned random amounts of time to wait before trying again. Without randomness, the messages might come back at the same time over and over again, forever after. Randomness is the grease that allows multiple Internet messages to slide by each other.

Randomness in computers is even more obvious in computer games. Imagine how boring those games would be if the bad guy always arrived at the same time, the space aliens always moved in the same pattern, and the virtual basketball player always passed in the same direction. It is the computer's randomness that makes the games come alive, that gives them personality, that makes them fun.

In reality, computers are unable to create true randomness. Instead, they just fake randomness through the use of pseudorandom numbers. These are sequences of numbers obtained through complicated arithmetic formulas (usually involving multiplying by large numbers, adding constants, and taking remainders upon dividing by a large power of 2). The resulting sequences are not truly random, but they are so mixed up and unpredictable that they appear to be random for all intents and purposes. The design and study of pseudorandom number generators is a major area of research, and every generator has certain flaws that make it not quite truly random, but hopefully good enough for the problem at hand. In short, computer designers work very hard not to avoid randomness, but to create it.

Monte Carlo Magic

During World War II, the Manhattan Project was set up in secret in Los Alamos, New Mexico, to work on designing and building the world's first atomic bomb. One difficult question was how much purified uranium was required before the bomb would explode. Calculating this critical mass was crucial to the success of the project. If it was underestimated, insufficient amounts of uranium would be produced and the bomb wouldn't work. Overestimating would be even worse: the bomb might go off prematurely, before detonation time, killing the wrong people in the wrong place.

The actual mechanics of the atomic-bomb chain reaction—neutrons causing atoms to split and release energy (according to Einstein's famous formula, $E = mc^2$), while in turn producing more neutrons to continue the process—are extremely complicated. Even the great Manhattan Project scientists could not calculate the resulting critical mass theoretically. Instead, they set up some of the world's first computers, complete with punch cards carefully handled by human assistants. These computers were designed to simulate randomly the chain reaction and the motion of the neutrons. By repeating these simulations over and over again, the scientists got an increasingly accurate sense of how the neutrons in an atomic bomb would behave on average, and what fraction would escape. In the end, the Manhattan Project scientists correctly calculated a critical mass of about 15 kilograms. Atomic bombs were manufactured based on this calculation, and they worked just as predicted.

These primitive computer simulations were the world's first use of the *Monte Carlo* sampling method, which consists of repeated randomized simulations to approximate quantities that are too difficult to compute directly. (The name "Monte Carlo" was suggested the following year by the Polish mathematician Stanislaw Ulam, in reference to the famous casino in Monaco.)

The atomic bomb changed the world forever (and not necessarily for the better). But in its own way, the introduction of Monte Carlo sampling methods changed the world, too. With modern high-speed computers, randomized simulations can now be easily run from any office, often in the blink of an eye. They are used routinely by scientists, engineers, medical researchers, and statisticians to estimate the value of huge and complicated sums, integrals, and probabilities. They are especially useful in high dimensions, when many different quantities interact and must be handled together. For example, they help us figure out the likely effects of medical treatments involving many variables, and of detailed engineering designs for everything from buildings to rocket ships. Monte Carlo simulations are an essential tool of virtually every modern science.

Raisins and Chocolate Chips

Marjorie always makes the most delicious raisin chocolate chip cookies. Your kids always seem to be hanging out at her place, hoping for a free sample. You start to feel jealous and want to match Marjorie's culinary triumph. But you don't know her recipe, and she isn't about to tell.

After some experimentation, you decide that the ratio of raisins to chocolate chips is the key. Does Marjorie use the same number of raisins as chocolate chips? Or twice as many raisins? Or three times as many chocolate chips? If only you knew this one detail, you are confident that you could fill in the rest of the recipe, and regain your children's admiration.

Slowly you devise a plan. With a combination of bribes and threats, you convince your young daughter to bring you a few of Marjorie's famous delights. Rather than eat them, you carefully pick them apart and begin counting.

The first cookie has 11 chocolate chips, and six raisins. The second cookie has 14 chocolate chips, and eight raisins. The third cookie has nine chocolate chips, and four raisins. You begin to see the pattern: Each cookie has just about twice as many chocolate chips as raisins. Your Monte Carlo experiment has allowed you to estimate the previously elusive ratio: two chocolate chips for every one raisin.

That settles it. Excitedly, you pull out your ingredients, and get to work. You carefully count out a total of 200 chocolate chips and 100 raisins. You combine them with the other ingredients, bake to perfection, and serve. Your children's smiles are all the proof you need that your Monte Carlo experiment has been a success. You are bursting with pride, your family honor is restored, and Marjorie doesn't know what hit her.

Perhaps the earliest example of a Monte Carlo experiment is a clever scheme proposed in the eighteenth century by Georges Louis Leclerc, Comte de Buffon. His *Buffon's needle* experiment requires a large sheet of lined paper (or a striped floor), together with a needle (or pencil) that is precisely the length between consecutive lines. If you throw the needle or pencil at random and wait for it to stop moving, what is the probability that it will touch one of the lines? The surprising answer is that this probability is equal to 2 divided by pi (π). Pi is the famous and mysterious mathematical constant equal to the ratio of the circumference of a circle to its diameter.

This unexpected result means that your needle and paper can be used to perform a Monte Carlo experiment, for the purposes of estimating π. Simply throw the needle many times. Then, take the number of times you threw the needle, multiply that by 2, and divide the result by the number of times the needle lay across a line. The result should be pretty close to the true value of π, which is known by high-speed computer calculations to be equal to 3.14159265. . . .

In 1864, an American civil war captain, O.C. Fox, was recovering from battlefield wounds. As a distraction, he tried the Buffon's needle experiment. He threw his needle a total of 1,620 times, and obtained three different estimates of π: 3.1780, 3.1423, and 3.1416. Not bad at all. Of course, π can be computed much more efficiently with modern computers than with needles and paper. But the early Monte Carlo spirit of Buffon and Fox lives on.

Before moving on, I feel compelled to mention one particular variant of Monte Carlo experiments, called *Markov chain Monte Carlo*. In this form of Monte Carlo (which happens to be my research specialty), the experiment does not start fresh each time, like throwing the needle anew, or getting hold of a different cookie. Rather, each new experiment develops from where the previous experiment left off.

For example, suppose you wanted to measure the average pollution level in a large wilderness lake system. You might proceed by setting out in a canoe, and paddling this way and that way, from inlet to inlet and lake to lake, throughout the park, without any particular destination. Every five minutes, you might take a water sample and measure the pollutants. Each new sample would be taken just a short distance from the previous sample, thus continuing from where the previous sample left off. Still, if you averaged the pollution levels in many different samples, over many days of paddling, eventually you would get an accurate picture of the lake system. In fact, you would be running a Markov chain Monte Carlo algorithm. Modern computer programs use this same basic idea to do all sorts of calculations, in physics and biology and medicine and social sciences, that would all be impossible without the randomness of Markov chain Monte Carlo algorithms.

Randomness and Fairness

In 2003, the state of California went through a well-publicized and unprecedented election campaign to recall current governor Gray Davis and elect a new governor, Arnold Schwarzenegger. There were well over 100 candidates—including several actors—standing for the election, which many described as a "circus."

One issue was how to prepare the ballots. With so many candidates, the order of names was important. It was argued that the traditional solution of always listing the candidates in alphabetical order was unfair to those whose names come later in the alphabet. Instead, the California elections officials decided to use a randomized alphabet. They wrote the 26 letters on cards and pulled them at random from a revolving canister, resulting in the following new alphabetical order: R, W, Q, O, J, M, V, A, H, B, S, G, Z, X, N, T, C, I, E, K, U, P, D, Y, F, L. Lucky for those whose names began with R; not so lucky for those whose names began with L. But because the order was determined *randomly*, it was deemed that everyone had been treated equally.

Randomness has long been used to achieve fairness, by taking decision making away from individuals. Coin flips are sometimes used to decide an election that ends in an exact tie. Beginning in 1970, the military draft of young American men during the Vietnam War was determined by where their birthdays fell in a randomly selected ordering of all 366 days of the year (including February 29). To avoid excessive line-ups, the Toronto International Film Festival divides all the initial ticket request forms into a number of large bins (43 of them, in 2004), and chooses randomly which bin's requests are considered first (bin #10, in 2004). Teachers sometimes use randomness to determine the order in which their students must make class presentations. Paradoxically, randomness seems very unfair when disease or terrorism strikes, but as a way of settling human affairs, it may be the fairest mechanism we've got.

Randomness also sometimes provides fairness in sports. Coin flips are sometimes used to decide which team gets the ball first, or home-field advantage in the playoffs, or the first draft pick. Everyone accepts the coin flip as a fair solution, whereas no one would accept a non-random solution (say, the league commissioner making his own personal choice) in the same way.

An interesting incident related to fairness in sports occurred during the 1996 Olympic Games in Atlanta. The men's 100-meter sprint that year is well remembered by Canadians as the race in which Donovan Bailey won gold, thereby eliminating the shame of Ben Johnson's 1988 disqualification for drug use. But to British enthusiasts, that same 100-meter sprint is remembered for an entirely different reason: veteran British track star Linford Christie was charged with two "false starts" and disqualified from the race, thus ending his sports career in frustration. (Christie was so upset about his disqualification that he refused to leave the track for a minute or two.)

A closer investigation reveals some suspicious details. Christie had not, in fact, left the starting block *before* the starter gun had sounded (the traditional meaning of "false start"). Rather, his offence was that he left the starting block less than one-tenth of one second *after* the gun had sounded. Olympics officials had previously decided that no one's reaction time is ever less than one-tenth of one second, so any runner beginning the race within the first one-tenth of one second must have "anticipated" the starter gun. Such anticipation is against the rules because it usually arises when a runner is late getting into position, thereby delaying the race and controlling, to his or her advantage, just when the starter gun will sound. Christie had left the starting block just 0.086 seconds after the gun was fired. But if the entire point of the Olympics is to push humanity's physical limits, Christie's supporters argued, isn't it possible that somewhere, someday, someone would manage to react in less than one-tenth of one second? Just perhaps?

Meanwhile, sports officials (in particular, the International Association of Athletics Federations) insisted that the 0.1 second rule was necessary to prevent the "anticipation problem." And thus the dilemma remained: How to prevent runners from anticipating the starter gun, while not penalizing them for extraordinarily fast reaction times?

To probabilists, the solution was simple: create a simple mechanical device that fires the starter gun at a random time. The referee would wait until all runners are ready, then activate the device. At that point, the device would beep once (to indicate to all that it had been activated), then wait a random amount of time (say, between two and six seconds) before automatically firing the gun. (Even better would be a random wait time chosen from an exponential distribution, which is a kind of randomness that is completely impossible to anticipate.) That way, no one could hope to anticipate the (random) start time. The anticipation problem would be gone, and there would be no need to have a 0.1-second rule at all. Leaving the starter block before the gun sounded would still be a false start, of course, but leaving any time after the gun had sounded—even just 0.086 seconds after—would be perfectly legal. Unfortunately, Olympic officials didn't ask probabilists for advice, and the rule still stands.

In competitions where randomness cannot be controlled, it is desirable to make sure that randomness affects everyone equally. In a sailing race, the direction and speed of the wind are huge and unpredictable factors, but if all teams race simultaneously (rather than in sequence), at least everyone has to deal with the same conditions. In duplicate bridge (as opposed to rubber bridge), everyone plays the exact same hands of cards, so that no partnership can benefit from simply getting lucky and being dealt more Aces. The cards are as random as ever, but the randomness is applied equally to all, resulting in perfect fairness.

Randomness can also be used to divide costs or benefits fairly in cases where they cannot feasibly be divided concretely. The story of King Solomon illustrates the absurdity of dividing a baby in half to settle a custody dispute. However, the method of flipping a coin to decide who gets a baby is at least plausible as a solution (though still not such a great idea!). In less extreme situations, a coin flip often provides an elegant resolution.

Dividing the Restaurant Bill

Lunch was good, but the service was slow, and now you're running late. Finally the bill arrives, $17.76. You and your associate agree that with tip $20 sounds right. You agree to split the bill 50–50 and pay $10 each. The only problem is, you each just have a $20 bill.

You try to ask for change, but the waiter is too busy for you to get his attention. Now you're getting edgy. You have a meeting at the office in 10 minutes.

You consider offering to pay this time, and your associate can pay next time. Unfortunately, your associate is about to embark on a five-year assignment in Antarctica, so there won't be any "next time" to collect on for quite a while.

Then you have an idea. You offer to flip a coin: heads you pay the full $20, tails your associate pays. Your associate agrees, fairness prevails, and you make your meeting on time.

The next time you're out with a group, try dividing the restaurant bill using randomness and coin flips, rather than waiting for change. It's just one more way that randomness and uncertainty, when used properly, can sometimes be to your advantage.

13

Evolution, Genes, and Viruses

Randomness in Biology

There is surely no greater wonder in the solar system than the fact that human beings exist. It is incredible that matter, following the basic laws of physics over a period of billions of years, could evolve from simple chemicals, to living cells, to primitive fish and reptiles, to larger mammals, to primates, and finally to human beings with all their sophistication and intelligence (well, most of the time). It almost defies comprehension. And without this amazing process of evolution, we wouldn't be here today.

Evolution requires genetic mutation and recombination. In this way, living creatures create offspring of slightly different varieties at each generation. Then, through a process of *natural selection* (sometimes called *survival of the fittest*), those offspring best suited to survive and reproduce themselves create additional, similar creatures. Over millions of years new species or subspecies develop.

The process of mutation and recombination, which produces the required diversity of offspring at each generation, is essentially a process of *randomization*. This randomness creates a wide variety of potential offspring, who then either survive and flourish or (more often) fail and die.

Insufficient randomness in genetic reproduction results in stagnation of the species and failure to evolve. On the other hand, too much randomness prevents a species from developing in a stable manner. So, for a species to successfully evolve and thrive, the amount of randomness has to be just right. Fortunately for us, this has indeed been the case, allowing humanity to evolve, very gradually, from the basic building blocks of the universe.

But how is this possible? Human DNA consists of about 3 billion chemical-base pairs. Each pair can be one of four kinds (denoted A, C, T, and G). Thus, the total number of possible DNA strands is 4 multiplied by itself 3 billion times—which equals the number formed by writing a 1 followed by 1,800,000,000 zeros. This number is so huge that it defies comprehension. Now, some of the human DNA base pairs are redundant, and could be of any type without affecting the human being at all. Many others vary from individual to individual, and thereby distinguish one human from another. However, a significant fraction of these base pairs have to be "just so" in order to create human life; any mistakes and we simply don't end up with a human being.

Purely random mutation could eventually produce a large sample of possible DNA strands. But it is mind-bogglingly unlikely, even over billions of years, that human DNA would be created suddenly, purely through random chance. So, how is it that humans are here at all?

The process of natural selection provides the answer. This process ensures that less fit offspring do not survive and reproduce. Thus, surviving offspring are weighted toward being more fit, more advanced, better able to live and flourish. In practical terms, this means that surviving offspring are weighted toward being more intelligent, more adaptable, and more cunning—in short, more like humans.

Consider an early, primitive life form, such as an amoeba, which reproduces over and over again. The Law of Large Numbers ensures that in the long run, each generation of this life form will, on average, be a little more advanced than the previous generation. Naturally, early generations will still be rather amoeba-like, and it will be a long time before anything more interesting emerges. However, over billions of years, and hundreds of millions of generations, the long-run march toward humanity (or some other sophisticated, intelligent life-form) seems almost inevitable. At some point, higher-order life (like us) will emerge.

So, in some sense, the reason that humans rule the Earth is the same reason that casinos get rich. In each case, the odds are weighted slightly in their favor, so they are bound to thrive in the long run. (Of course, how the *first* self-replicating life form arose on Earth is a whole other question. But once life appeared, natural selection took over.)

Processes similar to evolution can also be seen in quite different areas. For example, consider cuisine and diet. Over the years different styles of food become popular or unpopular. New foods are created or introduced to a society, and sometimes catch on and other times do not. This is a different form of "survival of the fittest," where food is "fit" if people like to eat it. It is even believed that many new foods are first created by accident; someone spills one food on another and is pleased at the unexpected combination. Once again, randomness is helping to produce new possibilities (whether species or foods), which can then be tested for fitness to survive. Without randomness, foods—like species—would be far less advanced and diverse.

Designer Blue Genes

"Like father, like son" is a simple expression of the fact that genetic information is handed down from parents to children. Predicting what a child will be like is very complicated, since different genes in different combinations affect different attributes. Furthermore, to some extent a child's traits and abilities are shaped by his environment and not by his genes. Still, the basic rule by which children acquire each *specific* gene is simplicity itself, and is based entirely on probabilities.

People's genes come in pairs (with a few exceptions). For example, there is essentially a single gene-pair that controls whether your eye color is Light (blue, or green, or hazel, or grey) or Dark (some shade of brown or black). (Recent evidence suggests that the situation is more complicated, and more than one gene may affect Light/Dark eye color. But for now let's assume it's a single gene-pair.)

If a person has a Light–Light gene-pair, then her eyes will be a Light color. Or, if she has a Dark–Dark gene-pair, then her eyes will be a Dark color. But suppose she has a Light–Dark mixed pair. In that case, her eyes will be a Dark color, because the Dark gene is *dominant,* while the Light gene is *recessive*.

If you see a Light-eyed person on the street, you can be sure that he has a pair of Light eye-color genes. But if you see a Dark-eyed person, he may have either a pure Dark–Dark pair, or a mixed Light–Dark pair; there is no way to tell which.

How do these genes get passed down to a child? Simple: the child takes one gene from each parent. And, the child has an equal probability of selecting either one of each parent's two genes, for a total of four equally likely possibilities. For example, if each parent has a Light–Dark pair, then one time in four the child will end up with a Light–Light pair, one time in four a Dark–Dark pair, and two times in four a Light–Dark pair. A chart called a *Punnett square* shows this pattern:

		Mother's Gene:	
		Light	**Dark**
Father's Gene:	**Light**	Light–Light	Light–Dark
	Dark	Light–Dark	Dark–Dark

Suppose now that two Light-eyed people have a child together. Since they are each Light-eyed, they must each have a Light–Light gene-pair. Their child then has no choice but to select one Light gene from each parent. The child will thus also have a Light–Light pair, and hence also have Light-colored eyes. In this model, if both parents have Light eyes, then so will their child.

On the other hand, suppose a Light-eyed mother and a Dark-eyed father have a child together. Then the mother must have a Light–Light pair, while the father could have either a Light–Dark or a Dark–Dark pair. If the father has a Dark–Dark pair, then the child has no choice: he will get a Light gene from the mother, and a Dark gene from the father, so he will end up with a Light–Dark pair, and thus Dark-colored eyes. On the other hand, if the father has a Light–Dark pair, half the time the child will get a Light gene from the father and end up with a Light–Light pair, and hence Light-colored eyes. So, if one parent has Light eyes, and the other Dark eyes, their child has probability between 0 and 1/2 of having Light eyes, and probability at least 1/2 of having Dark eyes.

Finally, suppose two Dark-eyed people have a child. In that case, each parent could be either Dark–Dark or Light–Dark, and we don't know which. If either parent is Dark–Dark, then the child must select at least one Dark gene, so the child will have Dark eyes. But if both parents are Light–Dark, there is one chance in four that the child will select the Light gene from both parents. So, if both parents have Dark eyes, then their child has probability at most 1/4 of having Light eyes, and probability at least 3/4 of having Dark eyes.

Sorting out the genes in a family involves some detective work. In my case, I have brown (Dark) eyes, as do both of my parents. On that basis alone, each of my parents could be either Light–Dark or Dark–Dark.

On the other hand, one of my brothers has hazel (Light) eyes. So, he must have a Light–Light gene-pair. How could this be? The only explanation is that my parents each have a Light–Dark pair, and my brother happened to get the Light gene from both parents. Now the pieces of the puzzle are falling into place.

Since my parents each have a Light–Dark pair, that means that each of their children will have probability 1/4 of having a Light–Light pair, and hence Light eyes. (In fact, one of their three children has Light eyes, which is about right.) They will also have probability 1/4 of having a Dark–Dark pair, and probability 1/4 + 1/4 = 1/2 of having a Light–Dark pair (since they could have gotten the Light gene from their mother and the Dark gene from their father, or the other way around).

What about me? Since my eyes are brown (Dark), I know I do not have a Light–Light pair. I could either have a Light–Dark pair or a Dark–Dark pair. A Light–Dark pair is twice as likely because I could have gotten that in two different ways (either a Light gene from my mother and a Dark gene from my father, or vice versa). It follows that I have a 2/3 probability of having a Light–Dark pair, and a 1/3 probability of having a Dark–Dark pair—but I don't know which I actually have.

It all gets pretty complicated. But whether tracing the genetics of eye color, or of other, more complex traits like diseases and deformities, predictions and understanding all amount to computing probabilities.

Bye Bye, Blue Eyes

You see her gazing at you from the bar. Magnetic baby-blue eyes that hold you transfixed. A night of dancing, a few drinks, and you are hopelessly in love.

"Let's get married," she gasps, between swigs of cognac and passionate kisses. "We'll have six children. All strikingly beautiful, just like I am!"

It sounds tempting. A boatload of brats with that same baby-blue eye color sure would take the world by storm. Such magnetism, such beauty—there would be nothing your children couldn't accomplish.

But then you remember. Your eyes are brown, as are all the eyes in your entire extended family, for many generations. You're virtually positive that you haven't got any Light eye genes inside you. And, since Dark eye genes are dominant, your children will all have Dark eyes, too. You might one day have grandchildren with blue eyes, but as far as your kids go, there won't be a blue eye—either wet or dry—in sight.

"Sorry, sweetheart," you say coldly. "It wouldn't work out. It's no good between us." Leaving her broken-hearted, you stagger home, to sleep it off.

Infectious Disease

Viral infections are a major concern to all of us. From the latest flu, to the worldwide spread of AIDS, viruses cause much pain, suffering, and death. The origins of new viruses are somewhat mysterious, perhaps involving genetic mutation, or transmission from animals, or even laboratory experimentation gone awry. But once a virus is upon us, the concerns focus on how widely it will spread and how many people it will affect. Epidemiology, the study of disease transmission, is at heart a study of probabilities.

Each person who is infected by a virus will eventually be either cured or die. From the patient's point of view, being cured and dying are polar opposites. But from the virus's point of view, these two eventualities amount to the same thing. Either way, the individual is no longer contagious, and thus no longer able to infect others. The virus, if it is to survive, must move on to other hosts.

So, from the virus's point of view, the only way to thrive is to keep infecting new people. Each time a new person is infected, the number of virus hosts increases by one. However, each time an infected person gets cured or dies, the number of virus hosts decreases by one. (Probabilists refer to such a system as a "branching process." It can be represented visually as a sort of tree, with the branches pointing from each individual to the people they infect.) This is the virus's eternal battle: trying to increase the number of people infected and decrease the number of people cured.

Viruses are clever; they know the Law of Large Numbers. They know that in the long run, the only important question is whether the number of virus hosts is increasing or decreasing on average. The virus will thrive only if this number is increasing on average.

How can a virus ensure this? Simple. It has to ensure that, on average, each infected individual in turn infects more than one other individual before being cured or dying. (If God told humanity to "go forth and multiply," then apparently He instructed viruses the same way.) So, the question of whether a virus will spread quickly or will die out boils down to one simple question: is the virus's reproductive number—the average number of people who are in turn infected by each individual carrying the virus—more or less than one?

Spread the Word

You have just heard that your brother Louie is in town tonight, and has agreed to come to dinner. How exciting! You'll have to gather the whole extended family together—the cousins, in-laws, and grandparents will all want to see Louie.

"Hey, Billy," you tell your son, "Uncle Louie is coming over for dinner tonight. Everyone is invited. Spread the word!"

Lazily Billy wanders outside and runs into your daughter, Sue. "Sue," he begins, "tell everyone to come for dinner tonight with Uncle Louie."

Billy wanders off looking for other people. However, he soon sees a salamander dart through the bushes, and he runs to catch it, completely forgetting his earlier mission.

Meanwhile, Sue is off to the library and doesn't give dinner or poor Uncle Louie a second thought. She tells no one.

Finally, it is dinnertime. Where you were expecting a huge crowd to greet your dear brother, you find no one other than your two children. It seems that the news of Louie's arrival didn't spread very far. Billy told just one person and Sue told no one, for an average of one half of one transmission per person. One half is much less than one, so the news quickly died out.

When it comes to the spread of infectious diseases, we are all tied together. Whether or not you get sick depends not just on how susceptible you are to infection, but also on how susceptible are all of those around you. For example, you might attempt to avoid colds by such practices as washing your hands often, minimizing contact with others and with shared objects like banisters and doorknobs, avoiding touching your own eyes or mouth, getting plenty of sleep, and so on. Such personal efforts may have a modest impact on your probability of catching a cold. On the other hand, if everyone in your neighborhood follows these same practices, fewer of your neighbors will get sick and you will be much better protected. And, if everyone in the world is similarly careful, each infected individual will, on average, infect less than one other individual. The virus will quickly die out and many people will be spared the discomfort and inconvenience of catching it.

Being protected from disease simply because those *around* you are protected is sometimes referred to as "herd immunity," and it is a powerful force in preventing the spread of illness. However, the idea of herd immunity sometimes tricks individuals into unwisely lowering their guard. A tragic example is the spread of HIV/AIDS, where many people assume they don't need to practice safe sex because their sexual partners "must be safe." This over-reliance on herd immunity has had disastrous consequences worldwide.

A similar issue arises with vaccines. If a vaccine is effective and everyone gets it, the disease will quickly die out. However, if everyone else besides you gets the vaccine, the disease will almost certainly die out *anyway*. Since vaccines are inconvenient at best, and potentially painful or harmful at worst, there is a temptation to let everyone else get the vaccine while you avoid the trouble.

A major vaccine controversy erupted in the United Kingdom when the government asked that all children be given an MMR (measles, mumps, and rubella) vaccine. A team of medical researchers led by Dr. Andrew Wakefield published a paper in *The Lancet* in 1998, alleging that this vaccine might contribute to autism. Central to their argument were eight cases of children developing symptoms of autism just a few days after being vaccinated. As a result, some parents stopped getting their children immunized, while others accused them of selfishly relying on herd immunity and compromising the overall effectiveness of the vaccine program. There was pressure to provide three separate vaccines (one each for measles, mumps, and rubella) to allow parents to select only the vaccines they wanted. The government refused, arguing that, for the sake of public health, all children should get all three components of the vaccine. Matters were not helped when Prime Minister Tony Blair refused to disclose whether or not his own son had received the MMR vaccine.

Subsequent studies did not confirm any link between the MMR vaccine and autism, and most researchers believe the MMR vaccine to be completely safe. Nevertheless, in the years following that 1998 paper, approximately 20% of parents in the U.K. did not get their children immunized, giving rise to fears that the diseases would spread widely. Indeed there is some evidence that incidence of measles in the U.K. has recently increased. The MMR crisis was a difficult situation that pitted personal freedom against herd immunity, parental love against trust of the medical profession, and individual caution against the collective good.

The most extreme method of controlling diseases is to completely isolate infected people from everyone else. Such methods have been applied to many diseases from leprosy to the bubonic plague (as potently described in Albert Camus's *La Peste*) to the SARS outbreak in China, Toronto, and elsewhere in 2003. Such isolation tries to prevent non-infected people from becoming infected. If it is successful, the number of new infections eventually decreases to zero, thus ending the disease.

Proper precautions and vaccines can greatly diminish the threat of disease; however, the fact remains that if a new virus emerges that is extremely contagious and also extremely fatal—like the one dramatized in the Dustin Hoffman movie *Outbreak*—it could wipe out a huge portion of humanity in a very short time. It happened with bubonic plague in the fourteenth century (which killed 25 million people in one five-year period); it happened with the influenza pandemic of 1918–1919 (which killed 25 million people in just one year); and for all we know it could happen again.

Replenish, Replenish, Replenish

Viruses aren't alone in worrying about replenishing their numbers; humans confront the same challenge. Each person produces a certain number of children before he or she dies—whether it is zero or one or two or some other number. In the long run, will the human population increase or decrease? Since two parents are required to create a child, the question is whether or not humans on average produce more or less than two children each. If they produce more than two, the population will increase; if they produce fewer than two, it will decrease. If each woman gives birth to precisely two children on average, in the long run the population will not change. (The United Nations defines the replacement rate as 2.1 children per woman; the extra 0.1 comes from not counting those women who die before adulthood.)

The answer is that the human population is increasing. Indeed, the world average is that each woman gives birth to about 2.65 children over the course of her lifetime. This *total fertility rate* is significantly more than 2, and it explains the huge increase in the world's population, from 2.6 billion in 1950 to about 6.3 billion today. (The United Nations now expects the total fertility rate to drop eventually to the replacement rate, so that the world's population will stabilize at around 9 billion in approximately the year 2300; in 300 years we will see if they are correct.)

On the other hand, many countries (especially industrialized ones) have a much lower total fertility rate. According to the *CIA World Factbook,* the average woman in Canada gives birth to just 1.61 children; in the United Kingdom, 1.66; in Australia, 1.76; and in France, 1.85. All of these countries require immigration in order to keep their populations from decreasing in the long run. The United States, with 2.07 children per woman, is close to maintaining its total population without immigration (though they accept many immigrants, as well). In each case, the population increase involves the same mathematical principles as a virus spreading disease.

In addition to viruses and humans, all species of animals and plants are self-replicating, and similarly will either increase or decrease in population, depending on their average rate of reproduction. But the analogy doesn't end there. Computer viruses are also self-replicating (which is why they are named after viruses in the first place). Just like biological viruses, computer viruses need to be passed on to more than one other recipient, on average, if they are to continue. Computer viruses take this lesson to heart: they are programmed to vigorously search e-mail address books, and send out virus copies to every address they can find. That way, even if only a small minority of computer users are foolish enough to run the virus program (usually by unwittingly clicking on an e-mail attachment), this minority will cause virus copies to be sent to so many others that, on average, more than one new user is infected each time.

Chain letters are also self-replicating: they attempt to copy themselves to another host and start over again. Like viruses, chain letters are bound by the Law of Large Numbers. A chain letter thrives if, on average, each recipient sends out more than one copy; if recipients send fewer than one copy, on average, the letter will quickly die out. This is why most chain letters request that you send on five or eight or 10 copies. If they request too many copies, the task will be too burdensome, and virtually no one will comply. On the other hand, if they request too few copies, even a moderately high response rate will be insufficient to keep the letter going. For example, suppose that a chain letter asks for two copies to be sent, and one recipient in three complies with this request. That works out to an average of only 2/3 copies per recipient. Since 2/3 is less than one, the letter will quickly die out. Or for a more extreme example, suppose a chain letter requests that just one copy be passed along. Even if only a few recipients do not comply, the average will be less than one, and the letter will not continue. I am willing to bet that you have never, in your life, received a chain letter requesting that just one copy be sent on.

The night before the 2004 U.S. presidential election, the Republican Party sent out a mass e-mail, in the name of George W. Bush, urging the president's supporters to vote the next day. Recipients were also requested to forward the e-mail to five people. The Republicans knew that if their chain letter was to be successful, it would have to replicate more than one copy on average. If they requested that just one or two copies be forwarded, many recipients would fail to comply and the letter would soon die out. By requesting that five copies be sent out, the Republicans created a chain letter that survived and flourished in those pre-election hours.

Even word-of-mouth news is governed by the Law of Large Numbers. There is an old television commercial featuring a woman who is so pleased with her choice of shampoo that she tells two friends, who each tell two more friends, "and so on, and so on." I don't remember what brand of shampoo they were advertising (nor have I ever felt a need to exclaim to others about my hair care products). But the commercial does indicate how even neighborhood rumors are spread through self-replication. If a rumor is interesting enough that a listener repeats it to more than one other person on average, the rumor will spread rapidly. If not, then it will languish. Since two is more than one, that shampoo euphoria was guaranteed to be carried from friend to friend, forever more.

From evolution to genetics, from viruses to chain letters, self-replication shows that a little randomness can go a long way. The development of a new species and the spread of information can be accomplished quite efficiently, with a few simple rules and an awful lot of probabilities.

14

That Wily Monty Hall

Finding Probabilities from Clues

In daily life we are always assessing various probabilities: What are the odds that I will be killed if I cross the road? That I will pass my examination at school? That the woman of my dreams returns my feelings?

Sometimes we get additional information that causes us to reassess our probabilities: we read in the newspaper that a certain street has lots of traffic accidents; we hear that our teacher is a tough grader; the woman of our dreams smiles back at us. Each new piece of information causes an immediate re-evaluation of our assessed probabilities: maybe the road is more dangerous than I thought; I have no chance of passing my examination; perhaps she loves me after all.

Unfortunately, we don't always re-evaluate our probabilities correctly. For example, suppose your quaint little town has been shaken to the core. After years of utter tranquility, suddenly there has been a rash of murders. Police have identified five crazy homicidal maniacs on the loose. Further investigation brings a new detail to light: of the five murderers on the loose, four of them are men with beards. This, in a town of 10,000 people, of whom just 400 are men with beards. "Beware of Bearded Men," scream the newspaper headlines. You were always suspicious of men with beards and this confirms it: men with beards are dangerous!

You try to remain calm and to go about your daily business. However, the next night you are walking home on your dark, deserted road, and you hear a noise behind you. Someone is there! Your heart beats a little faster. In desperation, you remember your Probability Perspective. There are 10,000 people in this town, only five of whom are murderers. So that random person behind you has only five chances in 10,000 of being a murderer, which is just 0.05%. You relax a little. But then you pass a streetlight, and get a good look at him. You are utterly shocked to see that he has a beard. Now you're really done for. Since four murderers out of five have beards, there must be four chances in five, or 80%, that this person is a murderer, right?

No, not right. If there is a total of 400 men with beards, of whom four are murderers, the probability that a randomly chosen man with a beard is a murderer is just four chances in 400, or 1%. This is a far cry from the 80% chance you originally estimated. So, yes, having a beard does increase the probability that he is a murderer; however, it only increases the probability from 0.05% to 1%, which is still a very small probability.

This example is representative of the misjudgements that we all make from time to time when we assess groups of individuals. Consider how frequently we observe some characteristic (positive or negative) in a few members of some particular racial, ethnic, national, religious, or gender group, and immediately assume that most of this group shares the same characteristics.

In the foregoing example, seeing that the mysterious stranger had a beard did indeed change the probability that he was a murderer. It just didn't change it very much. So the question is, when you get new evidence, how much should your estimation of probabilities change? The science of re-evaluating probabilities based on new evidence is called *conditional probability*. It is often quite subtle, yet it arises in many different contexts.

My Lupus Scare (A True Story)

I once had to do a conditional probability calculation that was extremely personal and intense. As a healthy 25-year-old, I went for a routine but thorough medical examination. A week later, I received a letter in the mail. It was three lines long. It said, "The test for lupus indicates that you may have it. However, about 5% of the normal population tests positive. Please contact our office to arrange a follow-up appointment." I was shocked; until that moment, I had always taken my good health for granted.

I had to wait three days for the follow-up appointment. I was then told that the doctor could send my blood for a more sophisticated, DNA-based analysis, to determine whether I actually had lupus. The problem was, they might not get around to performing the test for a couple of weeks.

I was terrified and had to find some way to cope with my anxiety. I turned to probability theory. I needed to know, was I one of the 5% of the normal population who tests positive even though they are perfectly healthy? Or was I one of the nearly 1% who actually has lupus?

I saw that this was a question about conditional probability. The question was, conditional on knowing that I tested positive for the lupus test, what was the conditional probability that I actually had lupus?

Stated this way, the answer was simple. The total fraction of the population which would test positive was the sum of the 1% with lupus, and the 5% without lupus who still test positive, for a total of 6%. Yet only 1% of the population actually has lupus. So the conditional probability that I had lupus, given that I had tested positive, was equal to 1% divided by 6%, or 1/6, or about 17%.

That's not so large. Naturally, I did not want to have even a 17% chance of having a serious disease like lupus, but 17% was much better than 100%. It was enough to convert my total panic into something more like great fear and worry—a big improvement.

Finally I received a phone message from my doctor. The DNA-based test had come back negative: I did not have lupus. In the meantime, probability theory had helped me to cope with a very difficult situation.

Sometimes confusion about conditional probability is easy to see. For example, Myla Goldberg's novel *Bee Season* describes a spelling bee with 151 contestants, numbered from 1 to 151. The mother of one contestant hopes for a two-digit number, because most years a two-digit number wins. Of course, in this example any number is equally likely to win. Since there are 90 two-digit numbers (all the numbers between 10 and 99), the probability is 90/151, or 59.6%, that a contestant with a two-digit number will win. However, even if you have a two-digit number, and even if you're sure that a two-digit number will win, your probability of winning is still just 1/90, or 1.1%. So, your overall probability of winning is the probability 90/151 that some two-digit number will win *multiplied* by the probability 1/90 that you will win if a two-digit number wins. This works out to one chance in 151—the same probability as for any other contestant.

Another example is airplane accidents versus automobile accidents. We have seen that many more people die in automobile accidents than in airplane accidents. So, even taking into account that many more people drive than fly, it is still much safer to travel by airplane than by automobile. However, in addition to all the fatal automobile accidents that occur, there are also a lot of non-fatal ones, whereas most plane crashes result in most of the passengers dying. What this means is that, to avoid fatalities overall, it is safer to travel by airplane than by automobile. However, if somehow you knew for sure that you were going to be in an accident, it would be safer to be in an automobile accident than an airplane accident. In other words, planes are safer than cars (they have fewer fatalities), but plane crashes are more dangerous than car crashes. That's conditional probability for you.

The Proportionality Principle

Often we are confronted by situations in which two different possibilities are *a priori* equally likely, but then we receive new information in favor of one of them. For example, suppose you are waiting for the shower, and your sister has been in there for 15 minutes. You feel your blood boiling over, and your hatred of your sister—never small to begin with—keeps increasing. There's just one problem. You actually have two sisters, Alice and Brenda, and you're not sure which one is showering.

This isn't good—if you're going to hate a sister, you at least want to know *whom* to hate.

Suppose both your sisters' bedroom doors are closed, and you can't call out because your other sister (whichever one that is) may be asleep. In short, there is no evidence either way about which sister is in the shower. You decide it is equally likely that Alice or Brenda is ruining your morning.

Then you hear a noise over the din of the shower. It's vaguely melodic, and you realize that whoever is in the shower is singing. Her joy only makes you angrier. But at the same time, it gives you an idea. Perhaps you can update your probabilities based on this new information (the singing), and get a better idea of which sister should bear your anger. You know that Alice loves to sing, and virtually always sings when she's in the shower. Brenda, on the other hand, is only a moderately enthusiastic singer, and sings during perhaps one quarter of her showers. So, once you hear the singing, you figure it is more likely that Alice is in the shower, rather than Brenda.

But how much more likely? You think again. If Alice is four times as likely as Brenda to sing, then a singing showerer must be four times as likely to be Alice as to be Brenda. Aha, you declare. The odds of Alice versus Brenda are now 4 to 1. Hence, there is a 4/5 chance that you should be angry at Alice, and just a 1/5 chance that you should be angry at Brenda. Fresh thoughts of how to ruin Alice's life fill your mind.

This example illustrates an important idea, which I call the *proportionality principle* (it's a special case of Bayes' rule). If different possibilities (like Alice or Brenda being in the shower) are equally likely to begin with, and new evidence (like the singing) arrives, you should update your probabilities in proportion to the corresponding chances (in this case, in proportion to the chances of Alice or Brenda singing).

Once we understand this principle, it is easy to apply. For example, suppose a sly operator named Quick-Hand Charlie calls out, "Say, mister. Step right up. Play the three-card thriller!" As you cautiously approach, he continues, "Look at these three cards. One is red on both sides, one is black on both sides, and one is red on one side and black on the other. Right?" You agree, and he goes on. "Now mix the cards all up in this giant bag here. Pick one card at random, and plunk it down on the table with a random side up."

Hesitantly, you swish the cards around in the bag, grab one of them, and put it quietly in front of him. The side you see is bright red. "Okay, that side is red, right?" Charlie says. "So I guess you didn't get the black–black card, eh? You got either the red–red card, or the red–black card—and it must be 50–50 whether or not the other side is red or black, right?"

Charlie is angling for a wager, and you're getting scared. Quickly you remember your Probability Perspective. Originally all three cards were equally likely to be selected. Your new evidence is that a randomly chosen side of the selected card is red. Aha, this is conditional probability.

You decide to use the proportionality principle. You reason that if you'd selected the red–red card, then a randomly chosen side would always be red. But if you'd selected the red–black card, then a randomly chosen side would only be red half the time. Since the red–red card is twice as likely to show red as the red–black card, the probability must now be two to one that you've selected the red–red card. That is, there is a 2/3 chance that the other side of the card is also red.

"Sorry, Charlie," you reply. "It's not 50–50 at all. The other side is red with probability two out of three." To illustrate your point, you flip over the card, revealing another red side on the back. Charlie starts to reply, but he realizes you are too smart for him. He moves on to the next sucker.

In this example, many people just don't believe that the probability of the other side being red is 2/3. If you're not convinced by the proportionality principle, here are two other ways to get the same answer. (Mathematicians are always happy when they can get the same answer by different methods—then it *must* be correct.)

One way is as follows. At the beginning of the game, you had one chance in three of selecting the red–black card. Seeing a red side changes many things, but it doesn't change this chance. No matter what color you see, you still have one chance in three of having chosen the red–black card. So, the probability is 1/3 that the other side is black, and 2/3 that the other side is red.

Still don't believe me? Then try the following explanation for the three-card thriller. There are three cards, each with two sides, for a total of six sides. What you do in the beginning is pick one of the six sides at random. Since it's red, it must be one of the three red sides, all equally likely. But two of those three red sides (the two sides of the red–red card) have a red side opposite, while just one of those three red sides (the red side of the red–black card) has a black side opposite. Thus, there are two chances in three that the opposite side is also red.

So there you have it. By the proportionality principle, or by remembering the chances at the beginning, or by counting sides instead of cards, we see that in Quick-Hand Charlie's three-card thriller, once you see that one side is red, there is a 2/3 chance that the other side is also red.

Easy, right? So now you're ready for the most famous (or infamous) of all conditional probability problems, the Monty Hall Problem.

The Monty Hall Problem

On September 9, 1990, Marilyn vos Savant presented a probability problem in her "Ask Marilyn" column of *Parade* magazine. The problem was the "Monty Hall problem," named after the host of the television program *Let's Make a Deal.*

The Monty Hall problem supposes that a new car is behind one of three doors. You pick one door (say, Door #1). The host then opens some other door (say, Door #3), showing no car (in fact, showing a goat instead). You are then given a choice: you can either stick with your original door (#1), or switch to the other unopened door (#2). If the car is behind the door you end up choosing, then you win the car, otherwise you don't.

The question is, should you open your original door, or should you switch? Is the car more likely to be behind Door #1 or behind Door #2? Most people assume that you are equally likely to win whether you switch or not, but vos Savant asserted that you were twice as likely to win if you switched.

Unexpectedly, vos Savant's column about this problem ignited a storm of controversy, and thousands of letters were received. Mathematicians from such institutions as George Mason University, the University of Florida, the University of Michigan, and Georgetown University wrote in to complain that vos Savant's answer was incorrect, saying that she "blew it big," that her "logic is in error," and that she is "utterly incorrect." One intemperate academic even wrote that vos Savant herself was the goat!

In response, vos Savant issued a challenge to "math classes across the country," asking them to try out the game as an experiment. She asked that they try not switching for 200 trials, and then try switching for the next 200 trials, to see which way they won more often. Many elementary-school math teachers took her up on this challenge, and their letters indicate great delight. "Our class, with unbridled enthusiasm, is proud to announce that our data support your position," said one. "Their joy is what makes teaching worthwhile," gushed another. "The results were thrilling!" exclaimed a third. (On the other hand, one student expressed pleasure merely because the experiment "got me out of fractions for two days.")

This Monty Hall problem struck a nerve with mathematicians, teachers, students, and the general public alike. How can we solve this tricky little puzzle?

The first thing to note is that the answer depends on the behavior pattern of the host. For example, suppose the host doesn't like you, and therefore offers you a chance to switch *only* when your original guess was correct. In that case, of course you should never switch, since the host allows you to switch only when switching would be a bad idea. Conversely, suppose the host really does like you, and therefore offers you a chance to switch *only* when your original guess was *not* correct, and switching is thus a good idea. In that case, of course you should *always* switch.

To remove these ambiguities, we will state our assumptions clearly. We assume that originally, before you made your selection, each of the three doors was equally likely to contain the car. We also assume that, whether your original selection was right or wrong, the host will *always* open a door that is different from your selected door, and that does not contain the car. The host will then always offer you a chance to switch to the other unopened door. Furthermore, if your original selection happens to be correct, the host is equally likely to choose either of the two other doors (neither of which contains a car in this case).

Now that the assumptions are clear, it is time to use some conditional probability. Originally, all three doors were equally likely to lead to the car. We now have some new evidence—namely, the host has opened Door #3, which has no car. We then ask, what is the updated probability that the car is behind Door #2?

The proportionality principle tells us that the conditional probability that Door #2 (say) contains the car is proportional to the probability of seeing the new evidence (i.e., of seeing the host open Door #3) if the car is actually behind Door #2. Now, if Door #2 actually contains the car, then after we guess Door #1, the host has no choice but to open Door #3—the only remaining door with no car. So, the host will *always* open Door #3. The probability is 1/1 (certainty) that the host will open Door #3 in this case.

On the other hand, if Door #1 actually contains the car (that is, if our original selection happened to be correct), then the host has a choice. Half the time he will open Door #2, and half the time he will open Door #3. So, in this case, only half the time will the host show us Door #3. The probability is 1/2 that the host will open Door #3 in this case.

We conclude that the chance of seeing the new evidence (the host opening Door #3) is twice as great if the car is behind Door #2, than if the car is behind Door #1. By the proportionality principle, if we guess Door #1 and then the host opens Door #3, it is *twice as likely* that the car is behind Door #2 as Door #1. In other words, if we switch from Door #1 to Door #2, we will increase our probability of winning the car to 2/3.

So, vos Savant was exactly right, and her mathematician critics were exactly wrong.

If you understand why there is a 2/3 chance that the car is behind Door #2, then you are ahead of most of the mathematicians who wrote in to *Parade* magazine. But if you don't, then take heart. Once again, there are other ways to get the same answer.

When you first begin the game, you have no idea where the car is. Thus, your original selection of Door #1 is purely random, and it has just 1/3 chance of being correct. You know that the host is then going to open some other door which contains no car, so when he does so, it doesn't surprise you at all. In particular, the host's action has *no effect at all* on the probability that your original selection was correct. That is, the chance that Door #1 is correct is the same as it was originally, namely 1/3. So, there is a 1/3 chance the car is behind Door #1, and therefore (by process of elimination) a 2/3 chance the car is behind Door #2. Once again, we see that vos Savant was correct.

If you're still not convinced, try this. Suppose that when the host has a choice of which door to open, he decides by flipping a coin. And pretend that even when he has no choice, he flips the coin anyway, even though it means nothing. In that case, we can write a table of all the possibilities, as shown in Table 14.1. The six rows in this table are all equally likely. Three of them (the second, third, and fourth rows) correspond to the host opening Door #3. And, in two of those three rows, the car is actually behind Door #2. This means that if the host opens Door #3, then there are two chances in three that the car is actually behind Door #2. The probability of winning by switching is 2/3, just as vos Savant said.

Table 14.1 Possible Monty Hall Outcomes (Where Your First Choice Is Door #1)

Your Choice	Car Location	Coin	Host Opens
#1	#1	Heads	#2
#1	#1	Tails	#3
#1	#2	Heads	#3
#1	#2	Tails	#3
#1	#3	Heads	#2
#1	#3	Tails	#2

So ends our tale of conditional probability. In summary, the proportionality principle tells us that if two possibilities start out equally likely and new evidence arrives, we should adjust our probabilities to be proportional to their probability of causing this new evidence. If Alice is four times as likely as Brenda to sing, then it is four times as likely (and hence has probability 4/5) that she is in the shower. Or, if a red–red card is twice as likely as a red–black card to show a red face, then it is twice as likely (and hence has probability 2/3) that the card is red–red. Or, if the host opening Door #3 is twice as likely to happen if the car is behind Door #2 than if it is behind Door #1, then it is twice as likely (and hence has probability 2/3) that the car is behind Door #2.

If you're still not convinced that the Monty Hall probability is 2/3, then remember that neither were many mathematicians the first time they heard this question. More important than this problem, or even the proportionality principle, is the fundamental fact of conditional probability: When new evidence arrives, you should update your probabilities accordingly—neither too much (as in the bearded-men scenario) nor too little (as for three-card thriller or the Monty Hall problem). If you keep that thinking in mind as you travel through life, you will make the appropriate inferences based on what you see and hear, and thereby end up a wiser person.

Scrappy Statisticians

Conditional probability is at the root of a serious, and at times heated, debate among statisticians. There are some statisticians who feel that many of the concepts of traditional statistical inference—p-values and margins of error and "19 times out of 20" and so on—are meaningless. These are statisticians who are "Bayesian"—pronounced BAY-zee-in, and named for Thomas Bayes (1702–1761), an English Nonconformist minister who developed many of the early rules of conditional probability. (Bayes died long before the debates in his name about statistical procedures, but Bayesian statisticians still regard him as a hero. He is buried in Bunhill Fields in downtown London, and the following inscription was added to his tomb 200 years after his death: "In recognition of Thomas Bayes's important works in probability this vault was restored in 1960 with contributions received from statisticians throughout the world.")

Bayesian statisticians believe that all uncertainties should be viewed in terms of conditional probability. For example, if they were testing a new medical drug, they wouldn't be content to know that the probability was less than 5% that the drug's benefits were the result of pure chance. Rather, they would want to know, given the results of the drug trial, what is the *conditional* probability that the drug is actually beneficial? Or, if presented with a poll, they wouldn't want to know the size of the margin of error. Rather, they would want to know, given the results of the poll, what is the *conditional* probability that their candidate will actually win the election?

To be more specific, consider the fictional Probalitus disease introduced earlier. This disease was normally fatal in 50% of cases, but a new drug has been given to, say, five patients and saved every one of them. Are these results sufficient to prove that the new drug is beneficial?

We have seen that classical (or "frequentist") statistics would have us compute the p-value—the probability of five patients in a row surviving by pure chance. This p-value equals 50% multiplied by itself five times, which is one chance in 32, or 3.1%. Since this p-value is less than 5%, classical statistics would conclude that those five patient recoveries could not have happened purely by chance, so the drug must be beneficial.

A Bayesian statistics analysis proceeds differently. It begins by assigning prior probabilities for the effect of the drug, representing what we think before we conduct the drug trial or see any evidence at all. Because we're not sure which is true, and because we want to be open-minded, we might declare our initial (prior) probabilities to be 50% that the drug is a miracle cure (and will save all patients), and 50% that the drug is useless (and does not change the disease's 50% survival rate at all).

Once these prior probabilities have been assigned, we next have to compute the conditional (or posterior) probability that the drug is beneficial, given that it saved five patients in a row.

What is this conditional probability? If the drug is a miracle cure, then any five patients will always survive. However, if the drug is useless, then the probability is just 1/32 of this happy event. Furthermore, we are assuming that according to our prior probabilities, the two possibilities (miracle cure or useless) were equally likely. So we can use the proportionality principle. It says that if five patients in a row survive, then it is 32 times more likely that the drug is a miracle cure than that it is useless. This means that the posterior probability that the drug is a miracle cure is 32/33 (96.97%), while the probability that it is useless is 1/33 (3.03%).

Thus, while a classical statistician would say that the p-value is 3.1% and thus conclude that the drug must be beneficial, a Bayesian statistician would instead declare that given the evidence, there is a 96.97% probability that the drug is beneficial. These two conclusions amount to pretty much the same thing: in both cases, there is strong evidence that the drug is beneficial. But whereas classical statistics would decide that the drug is helpful because the p-value is less than 5%, Bayesian statistics instead uses conditional probability to assign a final, "posterior" probability of 96.97% that the drug is helpful (and 3.03% that it is useless).

This dichotomy between classical and Bayesian statistics might sound like just a minor, technical difference, and in many ways it is. But to some statisticians, this difference is crucial to their philosophy of randomness. Most interesting is how passionately and personally certain statisticians take the distinction. Rather than refer to classical and Bayesian statistics as topics, they identify individual researchers as frequentists or Bayesians, and then criticize each other's approaches. Bayesians attack frequentists for the "contortions" of their "logically inconsistent" ideas, arguing that what we really care about isn't p-values or margins of error, but the actual probability that the drug is beneficial or that the candidate will win the election, conditional on the observed data. Frequentists counter that Bayesian statistics requires specifying your prior probability about what you believe before the experiment has even begun, and that there is no clear justification for how this prior probability is selected, thus rendering Bayesian statistics meaningless. This Bayesian versus frequentist debate was most passionate in the second half of the twentieth century, with each side publishing detailed attacks on the other, but it continues to this day.

I have learned the hard way that there are heated emotions on both sides of this debate, especially among the older generation of statisticians. I choose my words carefully in statistics department lounges whenever I am in "mixed" (Bayesian and frequentist) company, to avoid starting arguments. But if you ever find yourself with a large group of statisticians, and you want to cause a bit of trouble, ask them if they are Bayesian or frequentist. If your group happens to include a few passionate supporters of each side, then sit back and watch the fireworks.

15

Spam, Spam, Probability, and Spam

Blocking Unwanted E-mail

Conditional probability and Bayesian statistics have applications to many different areas of science and life. One that is becoming increasingly important is the blocking of unwanted commercial e-mail, or spam.

Spam (from "spiced ham") was originally a canned-meat product developed by the Hormel corporation in 1937, a follow-up to their "Hormel Flavor-Sealed Ham" product of 1926. During the fresh-meat shortages of World War II, Spam was distributed widely and consumed by soldiers and civilians alike in the United States, Canada, Great Britain, Russia, and elsewhere. It is estimated that over 5 billion cans of Spam have been consumed worldwide.

The 1970s British comedy group Monty Python later mocked the widespread availability of Spam. Their famous skit about a restaurant that offers such breakfast delicacies as "Spam, sausage, Spam, Spam, bacon, Spam, tomato, and Spam" helped to make the word *spam* synonymous with any item that is overly abundant.

So it was perhaps inevitable that, in the electronic age, *spam* would be used to describe the unwanted, unsolicited, commercial mass e-mail messages that we all receive and delete every day. These messages try to convince us to buy products, send money, visit commercial Web sites, or otherwise profit the sender.

Where once these spam messages were infrequent and possibly even amusing, they are now so widespread that they have become a significant drain on the productivity of everyone who uses e-mail in their daily lives. Indeed, it has been estimated that currently over 50% of all e-mail is spam, and many expect this figure to keep on rising. The total amount of effort spent worldwide to sift through and delete all those spam messages, while being careful not to accidentally delete any real e-mail, costs many billions of dollars—not to mention huge amounts of frustration and annoyance.

Most people now agree that spam e-mails are a serious problem and must be stopped. But how? The anti-spam battle is being waged on many fronts, from legislation to technology to individual habits. But it appears that the most promising anti-spam measures involve probability theory.

When inundated with spam e-mails, everyone's first reaction is, "Let's get these guys!" Many people would be quite happy if the police rounded up the senders of spam, and threw them into jail (with no Internet access) for many years. Unfortunately, this is not as easy as it sounds.

One problem with the "get the spammers!" approach is that not everyone agrees on precisely what does or does not constitute spam. Sometimes it's simple to draw the line: a loving note from your mother is not spam, but an unsolicited and unexpected invitation to visit a pornography Web site is. But what about, say, an e-mail from that nice Mr. Jones at the local hardware store, announcing to the neighborhood a discount on hammers this week? Or, an e-mail from an on-line store where you purchased a sweater last month, telling you about an even better sweater available this month? (Or, for professors like me, yet another mass e-mail announcement about some upcoming research conference that I have no interest in attending?) Sometimes the lines can get blurred.

A more fundamental problem is that most spammers remain hidden, sending their spam from various anonymous Internet accounts, often switching in a hurry between different Internet Service Providers (ISPs). Complaining directly to the ISP results, at best, in termination of the computer account in question, but it is easy enough for the spammers to open up another one. (Even determining which ISP is involved isn't so simple, since spammers often forge that information in their messages.) Furthermore, the international nature of the Internet means that spam e-mails can originate from any country, and apprehending spammers requires knowledge and enforcement of complex extradition and other international treaties.

Lawmakers have introduced legislation against sending spam. In the United States, New York Senator Chuck Schumer and other lawmakers have introduced a "CAN-SPAM Act," which came into effect on January 1, 2004. This act provides for jail terms of up to five years for those who, among other things, use multiple computer accounts under false identities to transmit "multiple commercial electronic mail messages." (Apparently, this careful wording was chosen to allow the politicians themselves to continue to use mass e-mail to solicit political donations.) This is a promising development, but will it be enforceable and make a significant dent in the amount of spam sent? Many people are skeptical.

In short, catching spammers can be very difficult. To make real headway in the anti-spam fight, we must use other methods.

It is worth asking why spammers bother to send spam. Some of them are just cranks. Others have been duped into paying spam-sending companies to send out spam for them, in the mistaken belief that they will make a quick buck. However, the majority of spammers send out their spam in a sincere effort to make a profit.

Spam solicitations typically receive something like 15 responses per 1 million spam messages sent. This is equivalent to a response rate of 0.0015%—extremely low by any standard. So how do spammers make any money?

The answer, of course, is that the cost of sending out spam is incredibly low. Indeed, the current "street price" of hiring a company to send out spam e-mails for you is approximately $1 per 10,000 e-mails sent—much, much lower than the cost of sending out paper mail. By spending about $100, spammers can send out 1 million spam e-mails, and get approximately 15 responses. Even if they make just, say, a $10 profit from each response, this still earns them $150, for a net profit of $50.

Of course, if no one ever responded to spam e-mail, the response rate would go down to zero. In this case, sending spam would never be profitable, and eventually the spammers would stop bothering. So, the simplest thing we can do to stamp out spam is to never make purchases in response to spam. Just say no!

Unfortunately, even though most people never respond to spam, it seems there will always be some tiny fraction of e-mail users who will do so. Thus, non-response alone will not be enough to stop the problem of spam.

A related issue is how spammers get your e-mail address in the first place. If they don't know your e-mail address, they can't bother you. So another method of avoiding spam is to try to keep your e-mail address secret. Some ISPs and on-line sales companies actually sell your e-mail addresses to spammers; obviously, you should never knowingly do business with (or in any way provide your e-mail address to) such companies.

Also, some spammers simply "guess" various e-mail addresses, by sending messages to common-sounding names at popular e-mail domains. Thus, it is best to use a hard-to-guess account name.

In addition, many computer viruses, once they successfully attack one computer, automatically find (and send e-mail to) all of the e-mail addresses stored in that computer's address list. The only way to avoid this problem is to avoid having any e-mail correspondents (or at least none who will ever accidentally allow a computer virus to take over their computer), a virtually impossible proposition.

But most spammers get their e-mail addresses through the use of "spam harvesters," which are computer programs that automatically search Web sites and Web directories all over the world looking for new e-mail addresses. The only way to foil such harvesters is to keep your e-mail address off all public Web sites (except perhaps as an image, which computer programs can't read).

Unfortunately, it is virtually impossible to keep your e-mail address off every Web site, every directory, and every commercial list, all the while avoiding any correspondents who will ever be attacked by a computer virus. Thus, address secrecy, while generally a good idea, is again not a perfect solution to the ever-growing spam problem.

Block That Spam

If we can't catch the spammers, can't put them out of business, and can't hide from them, then what is left? Recent efforts have focused on turning the power of the computer back on the spammers, by using technology to prevent spam e-mail messages from ever reaching your e-mail inbox.

The Dastardly Spammity Spam

You're relaxing at home, finally getting a moment's peace, when the doorbell rings. You open it to see a shifty-eyed man with a waxed mustache and thin black tie, carrying some large hoops.

"Pleased ta meet ya," he begins smoothly. "Spammity Spam's my name, Hula Hoops are my game. What d'ya say, mister? I've got fat ones, thin ones, green ones, blue ones—"

"Sorry, not interested," you interrupt, closing the door and returning to your couch.

A minute later the door opens, and Spammity Spam sticks his head in. "I don't think you've really considered all the benefits of Hula Hoops," he begins.

Outraged, you jump up, and slam and lock the door. At least *now* he can't come back.

A minute later the doorbell rings again, repeatedly, over and over. Frustrated, you cut the wire to disable the bell. Phew.

Then there is a knocking at the door. First quiet, then louder. Argh! Rushing down to your basement, you find a few old mattresses and lean them against your front door, thus drowning out the knocking. Taking a deep breath, you try to relax.

What's that noise upstairs? It's Spammity Spam coming in through the bedroom window. "Hula Hoops are a great stress reliever," he continues, as if he hadn't been interrupted. "Fun and entertaining, too!"

Furious, you rush upstairs, push Spammity Spam back out onto the balcony, and nail thick wooden boards over all your house's windows.

Then you slump down onto the floor, in a fortified, lonely house with no sunlight. Your depression only deepens when you hear a noise from the roof, and realize that Spammity Spam is trying to get in through the attic.

Meanwhile, your kindly neighbor has just brought over some flowers from her garden. She is at your front door, getting frustrated as she presses the doorbell and knocks on the door without success.

In theory, we should be able to use the power of the computer to block spam. That is, every time a new e-mail address arrives, it should be automatically sent to a computer program that determines whether it is a legitimate e-mail (and lets it through), or a spam message (which is then deleted, or returned to the sender, or stored in a special separate "spam mailbox" for later inspection). Indeed, many ISPs have already set up such programs for their users. But how do they work and how successful are they?

The problem of blocking spam may be rephrased as a *classification problem*: How can we write a computer program to decide whether a new e-mail message is spam or not? In recent years, many researchers have taken this problem quite seriously, and have worked hard at finding good solutions. Perhaps inevitably, e-mail messages that are not spam have come to be known as *ham*. So, the question is, can we create a computer program (called a "spam filter") which is clever enough to decide which e-mail messages are spam and which are ham?

As a first attempt, we might set up a computer program that scans each e-mail message for various words and patterns, and classifies as spam any message in which they occur. For example, many spam e-mail messages are from pharmaceutical resellers, attempting to sell the drug Viagra. (The manufacturer of Viagra, Pfizer Inc., is in no way responsible for these messages.) Aha, you think. To deal with this problem, I will program my computer to automatically block any e-mail message that contains the word *Viagra*. Victory is at hand.

However, several problems may arise. One is that your computer might create "false positive" spam identifications. This means that some perfectly legitimate e-mail messages might mistakenly be classified as spam. For example, perhaps a work colleague ends an otherwise serious and important message with a comment like "Sorry to be slow in sending you this information. First I had to delete seven spam e-mail messages trying to sell me Viagra," or a joke like "Apologies if this e-mail was too boring—at least it's better than another advertisement for Viagra," or advice like "Check out that new probability book; it has a fascinating discussion about e-mails that contain the word *Viagra*." Unfortunately, your spam filter—a computer program and therefore not very smart—would see the word *Viagra* appearing in the message and automatically classify the message as spam. This would prevent you from reading your colleague's entire message.

A second problem is that the spammers will attempt to get around your spam-blocking program. For example, they might decide to misspell the word *Viagra* slightly. Indeed, in the past month I have received spam messages urging me to purchase such products as "Via.gra," "viagara," "Viagkra," "V*iagra," "V|@gra," "Vl@gra," "Vi@gra," "V-l-A-G-R-A," and even "Via<alt=3Dlkfrujv>gra" (this last one because HTML-format mail readers won't display the part between angle-brackets). All of these spellings are easily recognized by a human as meaning "Viagra," but your spam-blocking computer program will probably miss them. More sophisticated spammers might not even use the word *Viagra* at all; they might instead just rename their product, and describe its supposedly wonderful effects in other ways, again fooling your computer program.

Even if you do somehow manage to stamp out all of the Viagra advertisements, there are still so many other types of spam. You can easily spend your whole day adding different words and different spellings to your computer program's list, while also worrying about false positives causing your real e-mail not to reach you.

It all seems pretty hopeless. However, once again, probability theory saves the day.

Spam Probabilities

Rather than trying to figure out all the words that spam e-mail might contain, and hoping that no legitimate e-mail messages contain them, a new approach is to get the computer to estimate the *probability* that each new e-mail message is spam.

Spammity Spam Revisited

Depressed at missing the neighbor's visit and longing for sunshine, you try a different approach. You take down all the wooden boards and mattresses, and reconnect your doorbell. Next to your couch you connect two buttons, one labelled "Spam" and the other labelled "Ham."

A few minutes later, the doorbell rings. You hope that it is your nice neighbor, so you reach toward the "Ham" button, to automatically open the door and let her in.

"Not so fast," you decide. Without rising from the couch, you start to think. Your visitor rang the doorbell twice, just the way Spammity Spam used to. Hmmm.

Just before the bell, you heard a clickety-clack very similar to the noise Spammity Spam's boots made climbing up your porch steps. Hmmmm.

Out the window, you can see a shadow, and while it's blurry and hard to make out, you think you might see a large circle mixed into the image, kind of like the shadow that a Hula Hoop might make. Hmmmmm.

Combining all of this evidence together, you compute that there is a 97% probability that your visitor is the dastardly Spammity Spam. That's a very high probability. Thus, without getting up, you press the "Spam" button. Immediately, a giant spring is released from under your porch, sending your visitor hurtling across the street.

"Phew," you think, relaxing comfortably on your couch.

At the same time, you feel just a little nervous. "I sure hope that wasn't my neighbor!"

Many of the newest spam-filtering computer programs operate, at least partially, by automatically estimating a probability that the incoming e-mail message is spam. If this probability is large enough (say, larger than 90%), the message is classified as spam.

At the beginning of their use, these programs require human beings to classify a large collection of spam e-mail messages as spam, and another large collection of real e-mail messages as ham. The programs first train themselves on these two collections. They do this simply by counting how many times each word appears in the spam messages, and how many times in the ham messages. For example, perhaps the word *Viagra* appears in 52 spam messages and in just one ham message. The computer program will then assign the word *Viagra* a high spam probability, perhaps around 98%. (The precise rules vary from program to program.)

The computer program has thus determined that if the word *Viagra* appears in an e-mail message, there is a 98% probability that the message is spam. Does this mean that any message containing *Viagra* must be spam? No. The program must also consider all the other words in the message.

Suppose that in addition to *Viagra,* the message contains lots of words appropriate to your company and/or to your typical e-mail usage (in my case, words like *statistics* and *research* and *lecture*; in some of my students' cases, perhaps *beer* and *grunge* and *party*). These words would occur often in the ham collection, and not so often in the spam collection. They would thus have very low spam probabilities, perhaps around 1%.

So what does the computer do? Faced with a message containing some words of high spam probability (like *Viagra*) and others of low spam probability (like *research*), the computer combines all of these probabilities into one "grand spam probability" that the message overall is spam rather than ham (sometimes referred to as the message's "spamicity" or "spam score"). It does this by conditional probability, using the Bayesian statistics approach. The computer assumes initially that each message has 50% prior probability of being either spam or ham. It then uses conditional probability to compute the message's true spam probability (spamicity), conditional on the words appearing in the message. In other words, a message's spamicity is simply the Bayesian posterior probability that the message is spam.

Finally, once the spamicity has been computed, the computer must decide whether to classify the message as spam or as ham. Typically this step is simplicity itself. If the spamicity is more than 90%, for example, the message is classified as spam, otherwise it is classified as ham. (Of course, the 90% threshold could be modified, perhaps to 80% [riskier] or 95% [safer]. However, the lower the threshold, the greater the risk of false positives, which should be avoided at all cost.)

There are, of course, many variations on this approach. And, some filters, such as *SpamAssassin,* combine word-based probability calculations with other factors, such as whether entire lines of the message are in capital letters (a perennial spammer favorite). But, in essence, the above description applies to the design of many of the new spam-filtering computer programs.

How well do spam filters work? For probability-based spam filters, everything depends on the collections of spam and ham e-mails that are used to train the programs and to assign spam probabilities to the various words. But if the collections are large enough, the computer program, which makes up in speed what it lacks in smarts, can discover patterns that humans might miss.

For example, Paul Graham writes in his oft-cited article "A Plan for Spam" that his probability-based spam filter has identified as having very high spam probability not only such obvious words as *promotion, guarantee,* and *sexy,* but also certain less obvious words like *republic* (from all the scam e-mails asking you to send money to faraway places like Nigeria), *madam* (from all those "Dear Sir or Madam" salutations), and *per* (from quoting prices like "$6 per package of 10"). Perhaps least obvious of all was *ff0000,* HTML code for the color "bright red," which turns out to have a very high spam probability, because spammers often use this code to give their messages extra emphasis.

Thus, an automatic probability-based program can discover patterns that a human might not have guessed. Even better, it is all done automatically, so that no one has to spend all day sifting through every word of every spam message that comes along.

An added side bonus concerns computer viruses that are spread through e-mail. From a filtering perspective, these virus e-mail messages are just like spam; they too can be added to your spam collection, so that your spam filter recognizes patterns in the computer viruses too, and learns to block the viruses along with spam.

Once your spam filter is working, it can continue to learn. For example, whenever the filter lets in a spam message, you can add that message to the spam collection yourself, thereby "telling" the computer program about a new kind of spam.

Over time, the computer should thus determine that, for example, *Vi@gra* should also have a high spam probability (even though the computer does not "understand" the connection between *Vi@gra* and *Viagra*). With luck, the computer should get better and better at figuring out which messages are spam, and which ones are ham. Indeed, Graham claims that he has developed a system for his own personal e-mail that filters out 99.5% of all spam messages, with less than 0.03% false positives, which is very impressive. (My own spam filtering isn't that successful, but over 80% of my spam is filtered out, which saves me a large amount of time and frustration.)

Procedures such as these, in which computers update their probabilities based on new examples, are sometimes referred to as "machine learning" or "artificial intelligence," and they are related to "Bayesian networks," or "neural networks." Indeed, such procedures are now being used for everything from detecting financial fraud, to identifying incoming missiles from military sensors, to improving search algorithms on the World Wide Web, to providing context-sensitive help to personal computer users. However, the computers aren't really "learning" anything, at least not in the sense that you or I learn. All they are doing is counting words and patterns, and computing the associated probabilities.

Your Spam or Mine?

One interesting question that arises with spam filters is whether the spam and ham collections should be developed jointly by everyone, or whether each individual user should develop his or her own collections.

At first glance, it seems that the collections should be joint. All for one and one for all. After all, if I want to avoid most messages containing the word *Viagra*, then probably you do, too.

But wait. Suppose you happen to be a biomedical researcher specializing in the physiology of reproduction. For you, the effects of Viagra might provide important evidence for your research. Thus, many of your ordinary e-mail messages might include the word *Viagra,* with no need for them to be filtered as spam.

Similarly, in my e-mail the word *probability* occurs often. If an e-mail to me contains that word, it is probably ham, with perhaps a spam probability of just 1%. However, that same word may occur less frequently in e-mails to a person who is less interested in probability theory. That doesn't mean that any e-mail to them containing the word *probability* must be spam, but it does mean that, in e-mails to them, this word is not as relevant to the spam versus ham decision. Just as one man's meat is another man's poison, and one man's terrorist is another man's freedom fighter, it seems that at times one person's spam is another person's ham.

Some modern spam filters (such as *bogofilter*) are designed to let each user develop his or her own collection of spam and ham e-mail messages, and to use different spam probabilities, depending on which user is receiving the e-mail. Others (such as *SpamAssassin*) instead use the same probabilities for everyone.

Each approach has its advantages and disadvantages. But I have noticed that using my own spam collection gives me a great psychological boost. When new spam arrives and gets through my spam filter, my annoyance is eclipsed by a feeling of power, as I transfer the message to my spam collection and update my spam probabilities. I can almost hear myself saying to the message, in a tough-guy police-officer voice, "You have been identified as spam. Anything you contain can and will be used against you, to update my spam probabilities and to block any further transmissions from you or your partners."

Let the Battle Begin

Spammers and spam filterers are in the midst of a fierce battle over the future of spam. If the spammers win, a larger and larger fraction of e-mail will be spam, to the point where one day e-mail might become virtually unusable, or else be restricted to our close associates who need a special password to send messages to us. If the filterers win, virtually all spam will be blocked, e-mail will work efficiently, and the spammers will give up in frustration. The theater for this epic battle is nothing more nor less than the spam probabilities as calculated by the various spam filters.

In an effort to subvert these probabilities, spammers—in addition to using more and more misspellings—are also adding more and more ordinary words to their spam. Indeed, in the past week I have received spam beginning with a random and incoherent assortment of such words as *petition, equatorial, decease,* and *obstinacy*. These words had absolutely nothing to do with the product being sold. However, they did manage to trick my spam filter into classifying the messages as ham.

Spammers are also trying to avoid using typical spam words. Rather than asking you to *buy* their product, and telling you how *cheap* it is, or how *sexy* their models are, spammers now sometimes use blander messages like "Hi, there, check out the following great Web site," followed by the URL to buy their product. These spam messages presumably have even lower response rates, but they are chosen in an effort to avoid spam filters.

Meanwhile, the filterers fight back by training with larger and larger collections of spam; by parsing the incoming messages more carefully (such as by distinguishing between words in the e-mail headers and words in the body of the message); by considering more and more features of the message (blank lines, non-standard punctuation, no name in the "From:" header, etc.). Eventually I suspect they will have to consider all the *pairs* of words; for example, *unique opportunity* sounds more spam-like than either *unique* or *opportunity* on its own. This adds an extra layer of complication, which some spam filters (such as *Spam-Probe*) are already trying to overcome.

And so the battle rages. Who will win? It's too early to say. However, understanding the battle at least makes it easier to figure out why spam messages look the way they do. In our worst moments of despair, as we delete spam after spam with no end in sight, at least we can take solace in the fact that this battle centers around our old friend, probability theory.

16

Ignorance, Chaos, and Quantum Mechanics

Causes of Randomness

Randomness is central to so many different aspects of our life—from bad things like cancer and terrorism, to good things like profitable investments, to fun things like rolling dice and dealing cards. But where does randomness come from? Are the individual components—the dice, the stock markets, the terrorists—actually random? Or do we just *think* of them as random, because we don't know any better?

For the most part, the randomness we experience is based on our own *ignorance*. If we just had enough facts and insights, randomness would disappear and we would be left with certainty. If we knew exactly how the dice were thrown, we would know how they will land. If we could read the mind of the terrorist, we would know where he plans to strike next. If we could see the planning documents of all the investors, we would know which stock prices will rise tomorrow.

The Restless Restaurateur

Your restaurant business is off to a rough start. The first Saturday, full of optimism, you hired four waiters and two cooks, but hardly any customers came and you wasted lots of money. The following Saturday, feeling more pessimistic, you hired just one waiter and one cook, and the place was so packed that you were unable to provide adequate service. Tomorrow is your third Saturday, and you need to make your hiring decisions now. What should you do? It all seems so random.

Discouraged, you go for a walk. A young couple is pointing at your restaurant, saying, "That place looks nice; let's take the kids there tomorrow." Farther down the street, a busload of tourists is checking into a large hotel and promising to explore the neighborhood tomorrow. On a lamppost, you see a sign announcing a meeting of a dining club tomorrow—at your restaurant. Then, in the newspaper, you are delighted to see that your restaurant has just received a positive review.

The pieces of the puzzle are falling into place. A few minutes earlier, your ignorance had caused your restaurant's prospects to seem so uncertain, so random. But with your newfound knowledge, there is much less randomness involved. You are virtually certain to have a full house tomorrow.

Pleased and confident, you hire a full staff complement. The next day, your restaurant is packed, your customers are satisfied, and the money keeps rolling in.

Randomness from Chaos

If randomness arises from ignorance, then where does ignorance arise from? Sometimes this is obvious. Who can possibly say where all the individuals in a city are planning to dine, or what the world's terrorists have up their sleeves, or what your child will be when she grows up. There are too many factors, too many unknowns, to do more than estimate probabilities and brace for randomness.

However, randomness also arises in situations that seem far less mysterious. Consider flipping a coin. The coin is of standard issue, it is right in front of you, you toss it in the air yourself, and then you catch it. There is no cloak and dagger, no hidden causes, no scheming enemies, nothing up your sleeve. What is not to know? And yet, when we flip a coin, all we can say for sure is that there is 50% probability of heads and 50% probability of tails.

The reason is that flipping a coin is an example of a *chaotic system.* This means that a very small change in how you flip the coin—pushing it a tiny bit harder or spinning it a fraction less—can have a large effect on the final outcome and change heads to tails. To know to a certainty whether the coin will come up heads or tails, you would have to know extremely precisely how hard you had pushed the coin, and how hard you had spun it. If you had a sophisticated laser-based measuring system, maybe, just maybe, you might be able to accurately predict the coin. But human eyesight is not that precise. We can see *approximately* how hard the coin was pushed, how fast it is moving, and so on, but our vision is not precise enough to make accurate predictions. Our ignorance about the coin, small though it may be, is still sufficient to make the final outcome completely random.

On the other hand, suppose you roll a ball along the floor toward a wall. In this case, based on the angle at which you initially pushed the ball, you can pretty much tell where it will strike the wall. If you change the ball's direction a tiny bit, the collision spot will move just a little bit. So rolling a ball on the floor is not a chaotic system. It is easy to predict, and a small amount of ignorance does not cause a large amount of randomness.

Physical systems can largely be divided into two groups. On one side are systems that are regular and unsurprising, not overly sensitive or chaotic, and that exhibit little randomness. These include everything from rolling a ball on the floor, to dropping a rock over a cliff, to the motion of the planets around the Sun. On the other side are systems that are very sensitive and therefore chaotic, and are correspondingly unpredictable and random. These include flipping coins, rolling dice, shuffling cards, and causing billiard balls to repeatedly collide and bounce off each other on a pool table. So, the next time you play poker, you can thank chaos theory for the fact that your opponent doesn't know what cards you hold.

The Chaotic Boyfriend

After eight months, you think you finally have your boyfriend figured out. He is happy when his work is going well, when he eats beef and drinks imported beer, and when the Red Sox win. He gets grouchy when he has problems at work, when he is served fish or milk, and when the Red Sox lose. It's all very simple.

One day, your boyfriend closes a big contract at work. That evening, you buy a case of German ale, cook up a pile of hamburgers, invite your boyfriend over, and settle down to watch the game. The Red Sox win 8–1. You look forward to a delightful, relaxing night together.

However, your boyfriend is still upset. It seems that the Yankees have also won tonight, so the Red Sox are still struggling to qualify for the playoffs. This one small fact has completely changed your boyfriend's mood, from relaxation and joy to anger and irritability.

Your boyfriend is definitely a chaotic system.

Chaos theory is also essential to the sequences of pseudorandom numbers that are used to simulate randomness on computers. In fact, these sequences are not random at all, but are based upon cold, hard, predictable equations. However, these equations are so chaotic—so sensitive to small changes—that the pseudorandom numbers jump around without apparent pattern, and thus seem to be random. Without chaos theory, there wouldn't be any Monte Carlo computer simulations, nor any random-seeming bad guys in computer games.

Our entire lives are governed by chaos, since small changes in the present can have a huge impact on the future. This property was nicely illustrated by the 1998 British movie *Sliding Doors,* starring Gwyneth Paltrow. Her character, Helen, is rushing to catch a subway train when a child briefly blocks her path. The movie presents two possible realities. In the first one, the child is quickly moved aside. The result is that Helen catches the train, meets a fellow passenger, returns home, and catches her boyfriend having an affair. In the second reality, Helen is delayed for a few extra seconds, misses the train, is stranded when subway service is later cancelled, gets mugged, and ends up in the hospital. Two completely different realities, and which one comes to pass depends solely on the triviality of how quickly a child is snatched out of the way on a subway station staircase. That's chaos theory in action. (In a nod to romance, Helen ultimately falls in love with the same fellow passenger in both realities. Oh well, no movie is perfect.)

Chaos theory provides a strong argument against the science-fiction concept of travelling backward in time. For example, in one of the original *Star Trek* stories, Dr. McCoy travels back to the 1930s and saves a woman from a car accident, thus ultimately allowing the Nazis to win World War II, which changes history and wipes out everything that we know as real. In response, Captain Kirk follows McCoy back and fixes things so the woman dies on schedule, thereby restoring everything to the way it was before. The problem is that, in the course of their escapades, Kirk and McCoy have interacted with many people, rented an apartment, worked for pay, made friends, taken up space, caused others to change their plans, and so on. (McCoy even caused the unintentional death of a homeless man.) Just as in the *Sliding Doors* example, any one of these small interactions would probably have a tremendous effect on subsequent events. With so many small changes having occurred, it is inconceivable that the world would be even remotely similar—much less identical—years after all those 1930s changes had taken place.

In *Back to the Future,* Michael J. Fox goes back in time and helps his dad to re-win his mother's heart. In the course of doing so, the dad acquires extra confidence. Then, in the present day, the family becomes more confident and successful than they were before. In *It's a Wonderful Life,* when James Stewart wishes he had never been born, his guardian angel shows him how different (and, as it happens, worse) the lives of his family and associates would have been without him. And, in the classic Ray Bradbury story "A Sound of Thunder," a dinosaur hunter travels 60 million years into the past and accidentally steps on a butterfly. The butterfly's future descendants—billions of them—are wiped out, leading to lack of food for other animals, and so on. After all that, what is the net change for modern society? A different person is elected president of the United States! So, in these stories, changes in the past do have a (slight) effect on the present. However, these effects are still not nearly as substantial as chaos theory tells us they would be, were such time travel possible. (I enjoy stories about travelling backward through time, but I am unable to completely suspend my disbelief, all because of chaos theory.)

Weather or Not

The most dramatic example of a chaotic system is the weather. Weather prediction can be something of an embarrassment to probabilists. Nearly everyone hears weather forecasts, complete with probabilities and percentages and satellite images, and nearly everyone observes that those weather forecasts are sometimes wrong. If meteorologists are ambassadors of probability, they are ambassadors that probabilists could sometimes live without.

There are many causes of this sad state of affairs. One is *observational bias*: people notice and remember incorrect forecasts much more than correct ones. In fact, weather forecasters get it right significantly more often than they get it wrong, with little gratitude in return. But the fact remains that, even with all of our modern computer models and satellite tracking and worldwide networks, humans are imperfect at predicting tomorrow's weather, and totally incapable of predicting anything more than a week ahead.

The reason is that, like coins and cards, weather is a chaotic system; very small changes today can cause significant differences tomorrow. We have all heard the tale of the "butterfly effect," originally proposed by American meteorologist Edward Lorenz (and later the basis for a 2004 Hollywood movie of the same name), whereby a butterfly flapping its wings in Brazil might cause, a few days later, a tornado in Texas. While some have claimed that this tale is an exaggeration, it nevertheless illustrates the fact that weather is the result of numerous reactions and collisions of unimaginable numbers of air molecules, water droplets, and other factors. It is completely impossible to keep track of them all, even with modern computers. These collisions do not run in simple patterns; rather they cause further collisions after further collisions, in highly unpredictable ways. Slight variations in current weather conditions can therefore lead to huge weather variations in the days ahead. So, even if our sensing equipment is extremely accurate, tiny errors or misjudgements will later lead to wildly inaccurate weather forecasts. The result of this chaos is that accurate weather prediction is simply too difficult a problem for our current science and technology to conquer.

Even judging weather prediction is complicated. If a forecaster says there is a 30% chance of rain tomorrow, and then it rains, does that mean he was wrong? Or 30% right? Or what? The fairest way to judge this is with what is called a *Brier score*: penalize the forecaster 30% times 30%, or 9%, if it doesn't rain tomorrow; penalize him 70% times 70%, or 49%, if it does rain. A typical forecaster might average a Brier penalty of 15% to 20%. This isn't too bad, but it isn't so great either. By comparison, a forecaster who predicted a 50% chance of rain every single day (regardless of the true weather) would receive a 25% Brier penalty, which is only a little bit worse. In fact, most weather prediction services don't share their prediction histories, so the public can't easily keep track of their accuracy (or lack thereof). And all the judging in the world will only confirm what we already know: weather forecasting is a very difficult science that is often right but is also often wrong.

So, the next time you are sitting in your car in the pouring rain while the radio is assuring you that the probability of precipitation is zero (yes, this has actually happened to me), try not to get angry. Try not to swear. Above all, don't blame the probabilists. It's not our fault—chaos is to blame!

True Randomness?

A core belief of traditional science was that randomness is caused purely by ignorance. From the time of Isaac Newton in the sixteenth century, physics has been governed by simple mathematical laws that tell us, in principle, precisely what will happen next. Based on the current position and velocity of a baseball, these laws can predict precisely how far it will sail and where it will land. These very same laws can predict the movement of Mars or Venus, days, months, or even years in advance.

Even for chaotic systems like the weather, where predictions are so difficult due to the system's extreme sensitivity, classical physics tells us that if we could measure exactly where each molecule of air and water is, and precisely how quickly it is moving, and if we had an infinite number of computers running for as long as we needed, then *in principle* we could predict the weather perfectly, too. Such predictions may be far beyond our current technology, but in theory they are just more complicated versions of flying baseballs and orbiting planets. A classical physicist's credo could be summed up as, "If we knew everything, then in principle we could predict the future precisely."

However, quantum mechanics changed all that. According to quantum mechanics, the universe works, on the most fundamental level, in terms not of fixed scientific certainties, but of probabilities and uncertainties. Quantum mechanics was developed in the early part of the twentieth century by such physicists as Max Born, Werner Heisenberg, Niels Bohr, and Erwin Schrödinger. This theory said, incredibly, that physics could no longer predict precisely what was going to happen; rather, all physics could determine were the probabilities that various outcomes would occur. For example, an electron orbiting an atomic nucleus could be in any of several different energy states, each with a certain specific probability.

The probabilities of quantum mechanics are given by a certain formula—the absolute square of the Schrödinger wave function—and this formula can be computed scientifically and accurately. But no matter how carefully you compute, no matter how many computers you have, and no matter how precisely you measure the current state of the electron (and of the rest of the universe, for that matter), you still cannot predict with certainty what will happen next—just the various probabilities.

Quantum mechanics goes even further. It provides a mathematical formula, the *Heisenberg uncertainty principle*, for how much uncertainty there will always be. The theory says that no matter how finely you measure a system, and no matter how much our technology advances, there will always be a certain pre-specified minimum level of uncertainty and randomness in everything that you observe or predict. Most maddeningly, it says that ignorance is not to blame. Even if you could repeat the exact same experiment, on the exact same materials, under the exact same conditions, nature's inherent randomness might cause an entirely different result.

These ideas shook science to its very core. For hundreds of years, science had been on a slow but steady march to understanding and predicting the universe more and more precisely. Where once we could only guess at how quickly sound travelled, or when the next eclipse would darken the sky, science had learned to calculate such matters with extreme accuracy. Quantum mechanics seemed to be putting on the brakes, to be telling science that its quest for precision was coming to a premature and unsatisfactory ending.

The idea that nature, at its most fundamental level, is inherently random goes entirely against our common sense. We are used to large, simple objects—like balls rolling across the floor—that follow clear patterns, continuing on in the direction they are rolling and bouncing off of walls in unsurprising ways. There is nothing random in their motions.

Quantum mechanics tells us that this is only because of the Law of Large Numbers. It says that a ball consists of billions of billions of molecules, each of which behaves randomly, but which taken together are entirely predictable. So, nature's randomness is not significant at sizes which we can see and experience, but it is fundamental at the unimaginably small scale of atoms and molecules. (Quantum mechanics also turns out to be important in certain large-scale astronomical phenomena, including the formation of black holes.)

Even if nature's randomness is confined, essentially, to the incredibly small, the question remains: How is this randomness generated? For example, suppose an electron has, according to quantum mechanics, probability 2/3 of being in a low-energy state, and probability 1/3 of being in a high-energy state. This means that if you use extremely powerful technology to measure the electron's orbit, there is a 2/3 probability of finding it in the first state, and a 1/3 probability of finding it in the second state. But who decides what state it is in? Does an all-powerful being watch over each electron, flipping coins and rolling dice to make choices with the right probability? Does the choice happen by magic?

The honest answer is that we do not know, despite nearly a century of using quantum mechanics. Indeed, the results of quantum mechanics are now used in much of modern technology, from microwave ovens to computer transistors to the futuristic-sounding idea of "quantum computing" (computers that actually make use of the laws of quantum mechanics to do faster computations). However, the mechanism of how quantum mechanics actually works remains mysterious.

The *many worlds theory*, first proposed by Princeton graduate student Hugh Everett, holds that each time quantum mechanics makes a decision involving randomness—for example, whether the electron is in its low-energy or high-energy state—it actually makes *both* choices, and creates two different universes, one for each of the two possible outcomes. According to this theory, rather than saying there is a 2/3 probability of the low-energy state, we should say that there is a 2/3 probability that you will end up in the universe corresponding to the low-energy state. Nature doesn't actually choose one of two possibilities; it chooses them both, in two different universes. Because other universes can't be detected by science, this theory can't be proved or disproved. However, it doesn't really solve the problem of who decides which outcome occurs—or alternatively, which of the multiple universes you end up in.

The inherent randomness of quantum mechanics is so counter to classical science that even many great scientists could not accept it. Albert Einstein himself—no stranger to revolutionary new ideas after proposing the theory of relativity, which revolutionized our understanding of time and space and gravity—disputed whether nature was truly random. In a letter to Born in 1926, Einstein wrote the famous words, "Jedenfalls bin ich überzeugt, dass der nicht würfelt" (I am convinced that He [God] does not play dice with the universe). Einstein (who was not religious) was indicating that the laws of nature must be described by precise, deterministic mathematics, which leaves no room for choice, uncertainty, or randomness. These laws may be very complex, and we may never understand them completely, but they should avoid any mention of probabilities or unknown factors. While Einstein did agree with many of the conclusions of quantum theory (and in fact, his explanation in 1905 of the photoelectric effect in some ways marked the beginnings of quantum mechanics), he never accepted its inherent randomness, its rolling of the dice.

The Grating Grader

You are disappointed to see that your essay about Shakespeare received a grade of C. Meanwhile, your best friend, Amy, who wrote a similar essay for the same teacher, got a B+. How could this be? Why did Mrs. Lane give Amy a better grade than you?

In desperation, you hire the detective Shady Shane to find out more. Shady quietly follows Mrs. Lane home, peeks in her window after dark, and watches her grade the next batch of essays. He then comes to you to report.

Here's what I saw, Shady explains. After glancing at each essay, Mrs. Lane picked up a six-sided die and rolled it on the coffee table. Upon viewing the result, she wrote a grade on the essay with a big red pen and set the paper aside. She repeated this exercise with all the essays, and she finished her grading remarkably quickly.

You start crying softly. You feel shocked, hurt, and betrayed. It cannot be, you insist. Shady Shane must be wrong, you wail. Mrs. Lane does not play dice with the grades!

Einstein's claim about God not playing dice became a rallying cry for those who opposed the philosophical underpinnings of quantum mechanics. These people insisted that there must be some "hidden variables"—tiny little instructions in the physical particles that told nature what choices to make. Perhaps these tiny instructions have not yet been found, they admitted, but someday they will be, and then nature's choices will be explained. The probabilities of quantum mechanics, like all other uncertainty, come from our ignorance—in this case, our ignorance about these tiny hidden instructions that we cannot see.

At a conference the following year, Bohr asserted that Einstein should stop telling God what to do. The dispute raged on, with great passion on both sides. A resolution of sorts came from the Irish physicist John Bell in the mid-1960s. Bell proved a mathematical theorem that demonstrated that the experimentally observed properties of elementary particles were inconsistent with the existence of local hidden variables—of tiny instructions telling nature what choices to make. Bell's theorem does not rule out the possibility of non-local hidden variables—that nature makes its choices based not on randomness, but on clearly defined rules based on other, faraway objects, somewhere else in the universe—but such an explanation seems even more counterintuitive than true randomness does.

Bell's work, combined with various physical experiments, has convinced most physicists that the randomness of quantum mechanics must be real, and that there are no hidden instructions. The debate still continues, and some scientists still hope for a non-random explanation of nature's behavior. But for the most part, science has accepted that nature, somehow, really does use randomness in making its fundamental choices.

Life in a Truly Random Universe

As a scientist, I am as uncomfortable as anyone with the randomness of quantum mechanics. But as a probabilist, I like it just fine. Above all, quantum mechanics says that probability theory isn't just a measure of our level of ignorance, it is also a fundamental law of nature. This makes understanding probability and uncertainty more important than ever.

The randomness of quantum mechanics also has some more practical benefits. Computer programmers need random numbers for everything from computer games to Monte Carlo experiments; however, they usually have to settle for pseudorandom numbers, which only fake randomness. But quantum mechanics allows us to use numbers which are truly random, straight from nature's own random choices. Indeed, there are various Web sites, such as HotBits, that provide free sequences of truly random numbers, as collected by Geiger counters measuring quantum mechanical radiation phenomena.

The use of such truly random number sequences in computer simulations is not yet widespread. The rate of generating randomness is too slow, and the probabilities of the resulting numbers are not always clear. However, these sequences do provide an exciting alternative to pseudorandom numbers. They also provide a connection to the true, genuine, indisputable randomness that is apparently a basic part of nature's inner workings.

17

Final Exam

Do You Have Probability Perspective?

Now that you are an expert at the Probability Perspective, you are ready to take your final exam.

* * * * *

1. You play tennis with your friend Dave. You know that he is not as good a player as you. However, today everything goes wrong: You slip in a puddle, several of your best shots go just inches long, two of his backhands dribble over the net, the sun is in your eyes, and you have a headache. As a result, you lose the match. Do you

   (a) throw in the towel, and never play tennis again.
   (b) despair that you have been cursed and will have rotten luck for the rest of your life.
   (c) arrange for Dave to have an "accident," which is the only way you will ever beat him.
   (d) suggest to Dave that you play tennis 10 more times, on 10 different days, because over the long run "luck factors" will cancel each other out.

2. You and your husband go out to dinner. Your husband brags that he can distinguish between Coke and Pepsi, and correctly identifies his current beverage as Pepsi. Do you

- **(a)** agree that your husband can distinguish between the two beverages.
- **(b)** admire your husband for his refined sense of taste.
- **(c)** suggest that your husband become a professional wine taster.
- **(d)** pour five glasses each of Coke and Pepsi, arrange them randomly, and see if your husband can correctly identify them all, thus making his p-value (the likelihood that he just got lucky) less than 5%, at which point you will finally believe in his special talent.

3. A salesman for Incredible Insurance, Inc. offers to insure your ukulele for you. He tells you that his company's insurance policies have such generous terms that it is always in the customer's interest to buy them. Do you

- **(a)** buy quickly, before he changes his mind.
- **(b)** give him a hug for being such a generous individual.
- **(c)** consider carefully and then decide that you should buy because insurance is the prudent choice.
- **(d)** note the company's huge profit margin, conclude that the company takes in more money than it pays out, and decide that you should buy the insurance only if the loss of your ukulele would cause you serious financial hardship.

4. Your shady dealings have caught up with you, and the head mobster issues his punishment. "One year from today, I will roll these 10 dice. If they each show a 5, we'll hunt you down and blast you to smithereens. Otherwise, we'll let you off." Do you

(a) spend the next year crying and quivering, with no hope for the future.
(b) kill yourself now, rather than postpone the inevitable.
(c) sign up for the space program, hoping you can hide from the mob on Mars.
(d) realize that the probability that 10 dice will all come up 5 is equal to one chance in six multiplied by itself 10 times, which is less than one chance in 60 million, so you really have nothing to worry about.

5. You're at a crowded party, with a plate of food in one hand and a glass of red wine in the other. A friendly businessman says hello and reaches to shake your hand. Your only hope is to balance your wine glass on a nearby windowsill, where you estimate it will have just a 5% chance of falling. Do you

(a) go ahead and balance the glass, satisfied that 95% of the time you will be fine.
(b) tempt fate further by leaving your glass on the windowsill for the next hour.
(c) for extra measure, bravely declare, "There ain't no way that baby is going to fall!"
(d) note that the carpet is white, so if your wine does spill it will have an extremely negative utility value, which even at 5% probability would more than cancel out the slight pleasure of shaking the businessman's hand, and decide instead to hold onto your glass and just smile.

6. The politician Sly Slade says you should support him because only he can combat the recent terrible rise in clogged kitchen sinks. Do you

(a) vote early and often for Mr. Slade.
(b) donate thousands of dollars to Mr. Slade's campaign.
(c) devote the rest of your life to railing against the evils of sink gunk.
(d) ask to first see an official list of the yearly rates of clogs per sink in your community, to determine whether or not they are truly increasing.

7. You buy a colorful new jacket and wear it for the first time. During the day you see numerous work colleagues and friends. Three of them comment on how nice your new jacket looks. Do you

(a) congratulate yourself on your great new purchase.
(b) buy several other similar jackets.
(c) go into business as a fashion consultant.
(d) reason that people usually keep negative reactions to themselves, so those three compliments are a biased sample, and that perhaps some of the other people who saw you had a more negative opinion.

8. Your husband said he'd be home for dinner by 6 o'clock and now it's nearly 6:30. Where could he be? Do you

(a) phone the police to put out a missing person's report.
(b) gather some friends together for a quiet commemoration of your husband's life and positive qualities.

(c) assume that your husband has been murdered and organize a posse to avenge his death by any means necessary.

(d) realize that the most likely explanation is that your husband is simply stuck in traffic, and watch a little television while you keep dinner warm.

9. You're playing poker, and determine that the only way for you to win is if your next card is the Ace of Spades. Do you

(a) close your eyes, click your heels, and think three times, "Ace of Spades please!"

(b) growl, glare, adjust your cowboy hat, and puff your cigar, to force the Ace of Spades to appear.

(c) make a list of all the Hollywood movies that ended with the hero being dealt an Ace of Spades when he needed it most.

(d) realize that all unseen cards are equally likely to appear next (so the probability of any one particular card is very low) and fold your hand before you lose your shirt.

10. You are hosting a dinner party scheduled to begin at 7 o'clock, and you are making a special cream sauce that needs to cook for exactly seven minutes and be served immediately. Do you

(a) start cooking the sauce at precisely 6:53.

(b) solemnly swear that you will serve dinner right at 7 o'clock, come what may.

(c) offer to commit hara-kiri if any guest's sauce is not cooked to perfection.

(d) recall that dinner-party arrival times have a large margin of error, so you had better delay and serve appetizers while waiting for the latecomers, and not cook the sauce until everyone is present.

11. You pick up a bunch of grapes and begin eating. You know that most of the grapes are delicious, but a few of them are sour and taste horrible. On the other hand, eating a sour grape together with some good grapes is not so bad. Do you

(a) hope that your bunch contains no sour grapes.
(b) throw the grapes out, because you can't bear the thought of eating a sour one.
(c) eat the grapes, slowly, one by one, vowing that if you hit a sour grape you will "take it like a man."
(d) eat the grapes three at a time, reasoning that the slight negative utility of one sour grape marring two good grapes pales in comparison to the large positive utility of not eating any sour grapes on their own.

12. In one year, the price of cars goes up by 8%, and the price of chocolate cake also goes up by 8%. Do you

(a) express astonishment that these prices are so closely related.
(b) launch an investigation into the underworld links between the car industry and the cake industry.
(c) speculate that cars are actually made out of cake.
(d) recall that correlation does not imply causation, and in fact neither of the two price increases causes the other; rather, both are caused by inflation.

13. A mutual fund advertises that their stock market investments made a tremendous profit three years ago. Do you

- **(a)** salute the fund manager as a financial genius.
- **(b)** immediately invest your life savings in the fund.
- **(c)** reach between your sofa cushions, to scrape up additional change to invest.
- **(d)** remember that any one year's profit might be just luck and demand information about the fund's performance in all recent years, to get a more accurate indicator of its true potential.

14. You see a performance by a self-declared psychic. He says he is feeling a supernatural connection, and asks if anyone on the first balcony has recently had a fight with an acquaintance whose name begins with J. A middle-aged woman puts up her hand, astonished, and admits that she had a serious argument with her son Jerome just last week. Do you

- **(a)** marvel at the psychic's incredible powers.
- **(b)** go out and buy all of the psychic's books.
- **(c)** hire the psychic to help you resolve all the conflict and confusion in your sordid life.
- **(d)** figure that out of the hundreds of spectators on the first balcony, and the many people that each of them knows, and the popularity of the initial J, and the propensity of humans to fight, it is not at all surprising that the psychic's prediction was true by pure chance alone, proving nothing.

15. Four months before your special day, you send out 300 wedding invitations to relatives and friends, near and far. You assume that about 150 to 200 people will accept. In the first week, you receive 27 responses, all in the affirmative. Do you

(a) congratulate yourself on your universal appeal.
(b) assume that all 300 invitees will accept and book a larger hall.
(c) anticipate throngs of admirers waiting outside, hoping to catch a glimpse of you as you leave the ceremony.
(d) realize that the responses so far are coming mostly from people who live nearby, and who are thus more likely to attend. So, these early responses constitute a biased sample, and the percentage of acceptances will surely decrease as more responses arrive from farther away.

* * * * *

As you have probably realized, the answer (d) is correct in every case. (Warning: The tests I give at university are not so simple.) Only answer (d) shows a true understanding of the principles and insights of probability theory.

If you answered (d) to all questions, then you, my friend, now have the knowledge. You have the power. You have the Probability Perspective. Use it well. Use it to understand the world more deeply. Use it to avoid fear. Use it to have fun. Use it to make better decisions.

The Probability Perspective will never replace all of your other critical thinking skills and decision-making methods—things like intuition and compassion and determination and honor and just plain common sense. But it will provide you with one more tool to better understand the world's randomness and your place within it.

Acknowledgements

It is a pleasure to thank the many people who made this book possible.

As a child, my interest in probability and mathematics was fostered by my family, including my parents, Helen and Peter; my brothers, Alan and Michael; and various grandparents, uncles, aunts, and cousins. Later on, my knowledge was greatly advanced by my wonderful mathematics teachers at Woburn Collegiate, the University of Toronto, and Harvard University, including my Ph.D. supervisor, Persi Diaconis; by a number of research collaborators, especially Gareth Roberts; by numerous supportive work colleagues, including my department chairs Gene Fabes, Mike Evans, Nancy Reid, and Keith Knight; and by interactions with many graduate and undergraduate students over many years.

My interest in communicating with the public was encouraged by several media interviews (including with Murray Campbell of the *Globe and Mail,* Leslie Roberts of *Global Television,* and Mary Ito of *TVOntario*), and also by reaction to the talks I gave to two University of Toronto alumni groups. In addition, I was fortunate to marry into a family of writers, journalists, editors, and television programmers, whose perspectives and insights were very helpful to me; I am particularly grateful for the generous assistance and unflagging encouragement of my stepmother-in-law, journalist Geraldine Sherman, without whom this book would never have been written.

I thank my editor, Jim Gifford; my agent, Beverley Slopen; and Kevin Hanson, Akka Janssen, Iris Tupholme, David Kent, Ian Coutts, Noelle Zitzer, Anne Holloway, and Roy Nicol, for believing in this project early on, and for providing loads of valuable assistance and advice as I travelled the long road from initial concept to final product.

On the technical side, I am indebted to the open-source computer community for providing free access to such powerful and reliable software as the GNU/Linux operating system, the C programming language, the R statistical analysis package, and the TeX mathematical formatting system. I am also grateful to the many librarians and staff members and agencies who helped me to find various statistical data.

Most importantly, I have benefited enormously from the tremendous emotional, practical, technical, intellectual, and editorial support of my wonderful wife, Margaret Fulford, to whom I dedicate this book. Without her I would accomplish far less and be far less happy.

About the Author

Jeffrey S. Rosenthal is a professor in the Department of Statistics at the University of Toronto. Born in Scarborough, Ontario, Canada in 1967, he was raised in a very mathematical family: both parents, one grandfather, and one uncle taught math, and another grandfather was an accountant. His interest in probability theory began at an early age; even as a child he enjoyed flipping coins, rolling dice, and computing probabilities.

Rosenthal received his BSc in Mathematics, Physics, and Computer Science from the University of Toronto at the age of 20; his Ph.D. in Mathematics from Harvard University at the age of 24; and tenure in the Department of Statistics at the University of Toronto at the age of 29. He has received several Natural Sciences and Engineering Research Council of Canada research grants, a Premier's Research Excellence Award, and research funding through the Mathematics of Information Technology and Complex Systems initiative, and is a Fellow of the Institute of Mathematical Statistics. A popular lecturer, he has won two teaching awards: a Harvard University Teaching Award in 1991, and an Arts and Science Outstanding Teaching Award at the University of Toronto in 1998.

Rosenthal has published two textbooks about probability theory, one at the graduate level (*A First Look at Rigorous Probability Theory,* World Scientific, 2000), and one at the undergraduate level (*Probability and Statistics: The Science of Uncertainty,* with M.J. Evans, W.H. Freeman,

2003). He has published over 50 research papers, many related to the field of Markov chain Monte Carlo randomized computer algorithms. He has also worked as a computer game programmer, musician, and improvisational comedy performer, and is fluent in French. He maintains the Web site www.probability.ca. Despite being born on Friday the thirteenth, Rosenthal has been a very fortunate person.

Index